From Research to Market

Alvaro Ossa

From Research to Market

A Comprehensive Guide

EDICIONES UC

Springer

Alvaro Ossa
Pontificia Universidad Católica de Chile
Santiago, Chile

ISBN 978-3-031-71339-2 ISBN 978-3-031-71340-8 (eBook)
https://doi.org/10.1007/978-3-031-71340-8

Jointly published with EdicionesUC

The original submitted manuscript has been translated into English. The translation was done using artificial intelligence. A subsequent revision was performed by the author(s) to further refine the work and to ensure that the translation is appropriate concerning content and scientific correctness. It may, however, read stylistically different from a conventional translation.

This Springer imprint is published by the registered company Springer Nature Switzerland AG
The registered company address is: Gewerbestrasse 11, 6330 Cham, Switzerland

Rich countries are so because they dedicate money to scientific and technological development, and poor countries remain because they do not. Science is not expensive; ignorance is.

Bernardo Houssay

I dedicate this book to Ofelia, whom I deeply admire, and to Olivia and Alicia, who teach me and help me discover this wonderful world every day. Together, we make each day a new adventure.

Foreword

Welcome to the Spanish language 'From research to market' knowledge manual in English.

I am fortunate enough to have worked closely with many Chilean colleagues since 2006 as they built up a very credible academic technology transfer system across the country. They are now about to embark on another major step forward by passing a national technology transfer law. Releasing this English translation of the original Manual in the Spanish language entitled: 'Del laboratorio al Mercado' is another step forward. This Spanish version also happened to be the basis for a very successful global reaching a Massive Open Online Course (MOOC) which since 2018 enrolled over 14,000 students in the three-weeklong course.

I believe that by making the From science to market Practice Manual available in English as well as Spanish, you are providing your country with a valuable tool by providing an even stronger foundation for moving innovations from Chilean universities to the marketplace.

The Spanish version allowed for the subtleties of legislative language and licensing operations in countries other than English-speaking ones. Moving from the original Spanish language manual to an English translation may well reveal interesting aspects of TT in other countries.

As outlined previously, the manual is structured to allow the practitioner to improve their skills in the mechanics of transactions. As the author, Alvaro says, it also includes real cases that show how technology transfer can have a positive impact on society and a region's economy. The text is divided into eight chapters, all addressing important aspects of the mechanics and the motivations of a having strong technology transfer activity.

This opens the door to explore how to 'Measure and Communicate the Value and Impact' of the academic technology transfer activities. The value can be measured in numerous ways, but the simplest is that such activities can lead to products that 'Save Lives,' 'Improve the Quality of Life' and 'Increase Competitiveness and Productivity of Companies.'

In the UK, researchers have organized the conversation of 'Impact of Research' into using standard language and categories, e.g., cultural; economic, societal; environmental; health; legal; political, educational and technological impact. In the late 1980s Angus Livingstone, then at the University of British Columbia, Canada published the table "Expanding the Metrics System for University Technology Transfer and Licensing" by using quantitative measures to measure: Academic impact; economic impact; financial impact, societal impact; spin-off company impact and finally the level of impact by technology field. For example, the measures for academic impact were number of students trained during technology development; number of publications based on the technology and impact rating of the journals; level of R&D funding stimulated by the technology and finally, the ability to further assist future innovation and academic progress.

One area that might be expanded in future editions would be 'Office Budget' and 'Office Structure.' For example, large US TTOs have staff that are generally structured into market areas, i.e., a life sciences team, a physical sciences team, etc., with a separate office-wide marketing team and leadership team. The members of each team bring specialized expertise to the team to the office and to the institution. Is this the best structure for different-sized offices?

For the budget, many US Offices separate the operations budget from the patent and IP Protection Budget. The first is for ongoing activities focused on building research relationships with the private sector and related (licensing and other) transactions, with money consumed on an ongoing, monthly basis. Part of this budget segment needs to be spent on continuing professional development for the staff. The second is the creation of valuable intellectual assets built up in an inventory over a longer time and not consumed on an ongoing monthly basis. This is a long-term asset of the institution, not the office.

Another additional area might be 'communication' and 'setting external and internal expectations.' Communications: Not only is it important to have effective outreach efforts on campus to alert researchers to the office services, but also to educate researchers about the mechanisms and timelines involved. For the individual researchers, engagement in commercialization activities can accelerate their research career by accessing new research problems, funding, equipment and partnerships. Expectations: It is important to have a communications activity to set expectations both internally and externally.

Internally: While individual projects can advance research careers and provide unique financing, TTOs are not cash cows. Leaderships of the institution need to understand that in the USA after 40 years of TTO activities, the percentage of commercialization deals that bring in more than $1 million cumulatively over the life of the license is about 2%. Again, in the USA, the percentage of original invention disclosures that are developed into marketed products is about 3%.

While these numbers are very small, AUTM statistics (gathered since 1992) show that over the first 25 years (to 2017), 124,000 licenses and options to licenses were signed by the 150–190 research institutions reporting numbers to AUTM. 46,000 of these were still active in 2017 and 20,500 have yielded some form of income—either upfront fees, maintenance fees or royalties. Each of the 124,000 agreements reflects a 1-to-1 partnership between a company and an academic institution. The benefits are not the big hits (rare), but the significant volume of partnerships formed over time—collaborations, additional financing, IP created, students hired, and institutional research bases deepened.

Externally: US studies have shown that larger cities with a research university have been able to build an innovation ecosystem to improve the local economy. In smaller communities, a commercialization activity at the local research university helps build the local economy and create jobs via spinout companies; serves as a magnet to attract significant research dollars from all levels of government and diversifies the local economy. The government dollars are spent in the local economy via salaries, supplies, etc. Other US studies have shown that vibrant cities based on one industry in the early 1900s were rejuvenated after that industry collapsed and disappeared by working with the local university to participate in the innovation economy. Two examples are Pittsburgh, PA. (after the steel industry left) and San Diego (after the US Navy downsized after WW2).

I personally believe that technology transfer (or knowledge exchange) is defined as the activity of transitioning campus creativity into corporate productivity to enhance the livability of our society in our lifetime.

I am excited by its potential to bring more Chilean practitioners into the international AUTM community of more than 3200 technology transfer professionals working in academic, research, government, legal and commercial settings. AUTM is dedicated to promoting and supporting technology transfer through education, advocacy, networking and communication initially in the USA, Canada and Europe, but now increasingly around the world as leaders realize that connecting selected academic results to the economy is a proven way to create an innovation ecosystem in the country.

Like Alvaro, I too hope that by reading this translation of his book, readers will help accelerate the global emerging activity of campus technology transfer bringing new innovative products to the marketplace thus saving lives, improving the quality of life and enhancing the productivity and competitiveness of our private sector.

John Fraser
Past President
and Chair—AUTM—2006–07
President
Burnside Development and Associates
LLC
Bethesda, MD, USA

Preface

During the global COVID-19 pandemic, science and technology, and above all the university-business connection, saved millions of lives worldwide. In many countries, researchers worked tirelessly to develop new vaccines and treatments that could fight the virus and prevent the spread of the disease. Collaboration between universities and businesses allowed these scientific developments to be quickly translated into products and services that could be used by doctors and patients worldwide. Thanks to these innovations, many lives were saved, and the impact of the pandemic on society and the economy was significantly reduced. This example shows how technology transfer can have a direct and positive impact on people's lives and society in general. Facilitating processes like this, which involve bringing science from laboratories to the market, transformed into products and services that improve people's quality of life, was the main motivation that led me to work in the world of technology transfer.

This book compiles experiences in technology transfer management from around the world. Through this collection, I seek to provide a practical and useful guide for those interested in developing successful technology transfer processes in their respective regions. Additionally, I include real cases that show how technology transfer can have a positive impact on society and a region's economy.

From Research to Market is divided into eight chapters that address the various stages of the technology transfer process. Each chapter can be reviewed individually, as well as read in a different order. Chapter 1 is an introduction to technology transfer. It includes several success stories and the experiences of the USA, Europe and Latin America. Chapter 2 explains what commercial validation is and how it is carried out. Chapter 3 addresses the key aspects of intellectual property, distinctions and protection mechanisms, including international norms and agreements. Chapter 4 describes the process of technology commercialization, types of contracts and the most commonly used negotiation mechanisms. Chapter 5 addresses the topic of science-based ventures, their characteristics and the conditions for their creation. Chapter 6, meanwhile, reviews financing alternatives according to the various stages of technological development and proposes a redefinition of the 3F model. Chapter 7 focuses on the process of technology transfer management, proposing two portfolio management models and explaining the concept of *technology marketing*. Finally, Chap. 8 discusses technology transfer offices, their roles and functions and performance metrics. It and proposes the six golden rules for the success of a technology transfer office in Latin America.

This book recounts some well-known successful cases of technologies developed in university laboratories and classrooms that made it to the market to solve specific problems, such as the isotonic drink Gatorade, created in the late 1960s at the University of Florida in the USA, or the cherry tomato variety, improved by two scientists from the Hebrew University of Jerusalem, who managed to create a sweeter and juicier fruit.

But in addition to these emblematic cases, *From Research to Market* tells other stories that have emerged in recent years and are examples that it is possible to promote technology transfer from different parts of the planet. This is the case of the anti-seismic devices created at the Pontifical Catholic University of Chile; the alternative therapy for the orphan disease called Morquio A syndrome, developed by researchers from the Pontifical Javeriana University in Colombia; a seawater desalination system or innovative nets that allow greater effectiveness in controlling fruit damage caused by the sun, both creations born at the University of Concepción. Also highlighted are the cases of a new assisted fertilization method, an invention that comes from the laboratories of the University of La Frontera; a bovine tuberculosis diagnostic system developed by scientists from the Austral University of Chile; or a thyroid cancer diagnostic test created by researchers from the Pontifical Catholic University of Chile. These are successful examples of technology transfers that have emerged from Latin America and are solving problems in different parts of the world. For all of the above to happen, promoting the 'triple helix' of innovation is essential through the articulation of academia, government and the private sector. The role of the government must be permanent and long-term. It is often mistakenly believed that only companies innovate and that the private sector is the fundamental link in the innovation ecosystem. The truth is that each sector has its mission, and only joint and articulated work will allow us to see medium and long-term results.

One aspect raised in the book is that when public policies are developed around the development of science and technology, a fundamental link in the chain is often postponed: the management of technology transfer. When seeking to advance science-based innovation decisively, it is necessary to professionalize the management of technology commercialization, establish processes, create incentives, train people and develop the necessary skills for all this to happen. This book is a guide for such professionalization and management.

From Research to Market is aimed at diverse audiences, including: technology managers and university authorities who are responsible for facilitating the academia-business linkage process; doctoral students and researchers who want to venture into the world of science-based innovation; scientific entrepreneurs interested in developing and launching their businesses; decision-makers in the fields of science and technology, to better understand technology transfer; and in general, anyone interested in science-based innovation as a driver of local and global development.

I sincerely hope this book helps to multiply the passion that drives us, allowing more actors to join this commitment to envision a future where science and innovation play leading roles and improve people's quality of life.

Santiago, Chile Alvaro Ossa

Contents

1 **Commercialization of Science and Technology** 1
Technology Transfer 1
Technology Transfer Mechanisms 3
Challenges and Barriers of Technology Transfer 4
Examples of Technology Transfer 7
Ideation Process 12
Technology Push Versus Demand Pull 12
Bootleg Research 13
Historical Context and Policies Supporting Technology Transfer 15
The United States 15
Europe 16
Latin America 17
References 18

2 **Commercial Validation** 21
Relevant Aspects of Commercial Validation 21
Market Opportunity 23
Problem and Solution 24
Value Proposition 25
Minimum Viable Product 26
Technical and Commercial Analysis 26
Development Status 27
Intellectual Property 28
Market Analysis 28
Regulatory Aspects 29
Financial Aspects 30
References 31

3 **Intellectual Property** 33
Key Aspects of Intellectual Property 33
Why Protect? 35

Classification of Intellectual Property ... 37
Copyright and Related Rights ... 37
Industrial Property ... 39
Protection of Computer Programs ... 44
Protection Strategies ... 45
References ... 46

4 **Technology Commercialization** ... 49
Key Aspects of Commercialization ... 50
Types of Agreement ... 51
Assignment Contract ... 51
License Agreement ... 51
License Option ... 52
Contract Negotiation ... 52
When to Commercialize? ... 52
Searching for Interested Parties ... 53
Preparation ... 53
Conditions of a License Agreement ... 54
Negotiation ... 60
Signing and Follow-Up ... 62
References ... 62

5 **Science-Based Entrepreneurship** ... 63
Key Aspects of Science-Based Entrepreneurship ... 63
Considerations of a Science-Based Entrepreneurship ... 65
Entrepreneur Profile ... 65
The Team ... 67
Business Development ... 68
Board of Directors ... 69
Shareholders' Agreement ... 70
Importance of the "*Exit*" ... 73
Enabling Conditions for the Creation of Science-Based Entrepreneurship ... 74
Entrepreneurial Capacities ... 74
Institutional Support ... 74
Public Policies ... 76
Financing ... 77
References ... 78

6 **Financing** ... 81
Financing Stages ... 82
Financing Mechanisms ... 82
Government Subsidy ... 83
From the 4F to the 3F + S ... 83
Angel Investors ... 84
Family Offices ... 85

Crowdfunding ... 86
Venture Capital ... 87
Corporate Venture Capital ... 88
Private Equity ... 89
Others ... 90
Risk Investment ... 91
The "Death Valley" ... 93
References ... 93

7 Technology Transfer Management ... 95
General Aspects of Technology Transfer Management ... 95
Technology Transfer Process ... 96
Science-Based Entrepreneurship Creation Process ... 99
Portfolio Management ... 100
Quadrant Methodology ... 102
Levels Methodology ... 105
Technology Maturity Levels ... 109
Technological Marketing ... 111
References ... 113

8 Technology Transfer Offices ... 115
Role of Universities in Technology Transfer ... 115
What and Why Is a Technology Transfer Office? ... 117
Functions of a Technology Transfer Office ... 119
Support Programs ... 120
Six Golden Rules for the Success of a Technology Transfer Office ... 123
The Commitment of the Authorities ... 123
Clear Rules ... 124
Incentives ... 127
Strategy and Structure ... 128
Technology Transfer Culture ... 129
Success Stories ... 131
Strategies for Strengthening a Technology Transfer Office ... 131
People Are an Essential Element for Achieving Technology Transfer ... 133
Performance Metrics ... 133
References ... 136

About the Author

Alvaro Ossa is the director of the Technology Transfer Office at the Pontifical Catholic University of Chile (UC), with over 20 years of professional experience in technology transfer and innovation. He holds a degree in Industrial Civil Engineering from the Federico Santa María Technical University in Chile and a Master's in Policy Studies in Education from University College London. In 2019, he was appointed Commissioner on the National Accreditation Commission for Chilean universities, recognized for his distinguished expertise in innovation.

Alvaro leads UC's research and development, intellectual property, and commercialization efforts. He also serves as the president of the Network of Technology Managers of Chilean Universities. A frequent speaker at national and international panels, Alvaro is also a columnist for several national press outlets and has taught various training programs in technological innovation management and entrepreneurship across the country.

In 2017, the Chilean government named Alvaro Technology Manager of the Year. From 2017 to 2021, the UC Technology Transfer Office was recognized as Chile's best technology transfer office. Under Alvaro's leadership, UC was Chile's top institution for patent applications from 2015 to 2023.

1 Commercialization of Science and Technology

Abstract

This chapter examines the crucial relevance of investment in science and technology as fundamental pillars for the development and well-being of a nation. Various challenges associated with technology transfer are explored, including the need to establish a close link between the academic and business sectors, as well as adopting a long-term perspective. To illustrate the practical application of technology transfer, real cases of successful technology commercialization from different parts of the world are presented. In addition, the "technology transfer" ideation model is examined in detail, encompassing approaches such as "technology push," "demand pull," and "bootleg research." Finally, a brief historical review of the development of technology transfer at the global level is carried out, highlighting milestones and significant advances in the field. This analysis allows us to understand the evolution of practices and policies in technology transfer, as well as identify lessons learned and potential areas for improvement. Together, this chapter provides a comprehensive and up-to-date view of the importance and progress of technology transfer in the global landscape.

Keywords

Technology transfer • Science and technology • University-business linkage • Technology push • Demand pull • Bootleg research • Innovation • Creativity

Technology Transfer

Research is an intellectual process, through which one can learn something new or even create things that do not exist, allowing for the improvement of the quality of education and even the improvement of people's lives through goods and services. It is a complex process with much uncertainty, and most of the time

A. Ossa, *From Research to Market*, https://doi.org/10.1007/978-3-031-71340-8_1

involves a long journey to achieve the desired result. It requires intensive work from researchers, and strong support from the institution where research is carried out, in both infrastructure and material resources, and administrative support. Resources are also needed, which mostly come from the government, to finance research to generate more and better knowledge that impacts society, either through the academic route (e.g., training people, scientific publications, or degree theses), as well as through the availability of research results that can be transformed into concrete solutions to real market and societal problems.

In this sense, there has been some controversy about the true purpose of scientific research. There are those who argue that science serves to better understand the environment around us, stimulating creative intellectual activity. On the other hand, there are those who believe that if certain scientific findings have potential applications and can solve specific societal problems, it is necessary to make every effort to make them so.

It is agreed that researchers seek to make an impact with the science they develop, either from the academic sphere or by solving a specific problem in society. The process of bringing a research result to the market is called technology transfer or "technology commercialization."

One of the most relevant aspects of the technology transfer process for a researcher is the satisfaction of contributing to society. It can be through the generation of a good or service that meets a need, such as developing a drug to cure a disease, carrying out a process that reduces industrial pollution, or creating a new method that improves teaching. In other words, the goal is to improve people's quality of life through scientific development.

The academic world has indicators and goals for academic progress, e.g., scientific publications. Technology transfer could also be an indicator of academic progress in a researcher's career if institutions recognized technology commercialization activity in their teaching evaluations. However, few institutions worldwide have implemented this.

From a financial point of view, the commercialization of science and technology could translate into a new source of funding for future research, as well as economic gain for researchers, depending on whether or not the strategy for transferring research results is for commercial purposes.

In this context, a challenge lies in ensuring that research results do not end up solely in scientific publications and their dissemination in the academic world, but that they reach the market and solve a real need for people. It is important to highlight that technology transfer was traditionally associated with engineering fields; later medical sciences were included. However, significant transfers have been seen in the areas of education, social sciences, and humanities. The commercialization of technologies is not exclusive to certain areas of knowledge but encompasses all disciplines that have a concrete application in the market.

Technology Transfer Mechanisms

For a research result to reach the market in the form of a good or service, it can be exploited directly or through the transfer to a third party for commercialization; it is rare for a scientific institution to directly market developments to end-users. In this sense, a research result can be transferred to an already established company, or a new company can be created to commercially exploit it.

Sell or Lease

Research results generate intellectual property that can be protected—an intellectual asset which can be transferred through two mechanisms: intellectual property rights can be assigned, transferring ownership to a third party, in which case the rights cease to belong to the original owner, or they can be licensed, which means that the original owner retains their property rights, and grants authorization for use, commercial exploitation, or massification to a third party under certain conditions.

As an illustrative example, let's assume that instead of having intellectual property rights, one has a house. This house can be sold, which would be equivalent to the assignment of intellectual property rights; or it can be rented, which is equivalent to licensing intellectual property rights. When selling a house, the new owner pays a certain price for it, and once the sale is completed, the house no longer belongs to the original owner, who loses all control over the property. The new owner can freely dispose of the house, paint it, rent it, sell it, or even demolish it. It may also happen that the new owner never lives in the house. In these cases, the previous owner has no influence over what the new owner decides to do.

Similarly, a research result can be licensed or assigned to another by the new owner. The new owner may also never use or exploit it, in which case the former owner cannot intervene.

Returning to the example, if the house is rented to a third party (equivalent to licensing a research result), the owner-landlord does not lose their ownership status; the tenant can only live in the house under certain conditions established in the rental contract. In the event of non-compliance with these conditions, the owner or landlord has the right to terminate the contract with the tenant. The same occurs when licensing a research result: the technology owner licenses or *rents* it under certain terms and conditions for use or commercial exploitation.

Just as renting a house may allow the tenant to sublet the property, the technology licensee may be granted the right to sublicense. In any case, this right must be granted with many safeguards, as it significantly impacts on multiple aspects of the license.

It is important to consider that the example presented helps to understand the difference between the assignment of rights and licensing. Of course, a different nature exists between a house and a research result: the latter is an intangible asset. The example does, however, allow for a quick and easy visualization of the difference between the assignment of rights and licensing.

Ultimately, the choice between assignment or licensing is a directly responds to each case's strategy. Among other aspects, it depends on the type of result obtained, its maturity level, the characteristics of its protection, what is intended to be achieved with its transfer, the type of financing available, and the interest of the party acquiring the technology.

The decision of whether to license or assign to an already established company or to promote the creation of a science-based venture to use or exploit a research result depends on an analysis of the particular case. It is advisable to create a new company when there are entrepreneurial capacities in the team, when it is a base technology that allows generating several technological developments from it, and when high capital investment is required.

Generating a science-based venture is not always the most suitable alternative; it requires a case-by-case analysis that takes multiple variables into account.

Challenges and Barriers of Technology Transfer

Technology transfer is a long and complex process, but with a potentially enormous economic and social impact. It also brings with it multiple challenges that need to be addressed.

On the other hand, there are multiple barriers to the development of technology transfer. Technology transfer is hindered by the existence of different challenges that are important to overcome, as they block interactions between universities and companies, thus preventing the development of innovations (Mazurkiewicz & Poteralska, 2017).

Understanding the Value of Technology Transfer

Various studies (Biggar Economics, 2017; Fraser, 2008) bring up the challenges that the technology transfer process must face. One of the biggest challenges is ensuring that countries know and understand its relevance; it is often believed that science and technology are the whim of a group of researchers who seek nothing more than personal prestige and advancement in their academic career. These researchers, however, want to make an impact with the science they develop through scientific publications and technology transfers that address society's needs.

Academia-Industry Relationship

Another important challenge is to achieve a greater university-company linkage to facilitate technology transfer, i.e., to link the development capacities of universities and research centers with the needs of companies and society. To ensure that science can solve real problems, it is necessary to dialogue with those who have these problems. This is achieved when both parties speak similar languages and approach positions; in other words, when universities learn to move at the pace of the industry, and the industry understands the importance of research and development (R&D)—its risks and costs (Fraser, 2008).

Long-Term Perspective

Few countries have a real and effective technological development strategy. Without a long-term State vision that truly involves the private sector, it is not possible to develop a robust science-based innovation system. On the other hand, the problem lies not only in the amounts to be invested, which are generally measured as a percentage of the nation's GDP, but also in who invests in R&D. While in developed countries, there is a balance between State and private contributions, and, in some cases, private contributions are even the majority, in less developed countries R&D is State financed, with the imaginable budgetary constraints.

On the other hand, it is very important that institutions have a commitment and long-term vision to develop applied research with impact in order to promote and strengthen their capacities to transfer their results. This should contribute to researchers becoming more interested in developing research that is oriented to the needs of the industry and has prospects for solving a specific problem from the outset (Fraser, 2008).

In addition to the above, Mazurkiewicz and Poteralska (2017) argue that in some cases, much time is required for technological development. This can lead to discouragement and disinterest from potential customers who want the technology to be developed as quickly as possible. Finally, it may also be that a prototype version of a technology does not work in the same way or is not compatible when industrial scaling is required, making its transfer completely unviable.

As studied by Mazurkiewicz and Poteralska (2017), time is among the economic and organizational barriers. While R&D organizations generally think about long-term developments, companies seek solutions in the shortest possible term; that is, they have a different view on the time needed to reach a solution.

Resources

It is key that governments invest in science and technology. Rich countries are so because they dedicate resources to scientific-technological development, and poor countries remain so because they do not: "Science is not expensive, ignorance is expensive" (Pigna & Pacheco, n.d.).

Scientific and technological development allows for the diversification of the productive matrix, making a nation more competitive and, above all, improving the quality of life for people. Countries must strive to understand, encourage, and invest in science and technology for growth and the well-being of the nation. It is important that they have a long-term State vision and not a limited short-term view (Mazurkiewicz & Poteralska, 2017).

In particular, one of the problems that arise in technological development through university-business links is related to who finances the developments. On the one hand, there is the interest of universities and R&D centers for companies to finance the entire process. That, however, is opposed to the interest of the industry, which is to finance the final result but not the research process itself.

On the other hand, sophisticated laboratories and equipment are required to achieve innovation based on science and technology. Further, it is not enough just to implement them initially; it is also necessary to maintain and update them; in order to generate world-class technological developments, it is essential to have world-class infrastructure (Mazurkiewicz & Poteralska, 2017).

Technological Development

Among the technological barriers is the intrinsic difficulty of developing any new technology (Mazurkiewicz & Poteralska, 2017). Since it is a high-risk process, it is not only necessary to prove that the technology works, but also to demonstrate its usefulness in solving a real problem. Sometimes technologies are too sophisticated, making it difficult or impossible to adapt them for use in the market. In this sense, it is much less risky to use already developed technologies than others in the process of development.

The authors argue that another factor considered a technological barrier is the limited technical capacity that may exist in research teams, which can lead to developing a single possible solution for the existing problem. This is a barrier, as it can lead to not seeing the problem globally, but rather focusing on a particular solution.

On the other hand, the fact that developments involve high technical requirements and personnel can make the cost very high and, therefore, often make their implementation and transfer complex.

The lack of an accurate assessment of the commercialization of technology and the absence of a plan for its implementation make up another constantly present barrier. R&D centers and universities focus heavily on advancing and disseminating knowledge (through scientific publications, for example) rather than patenting their findings; this focus deeply collides with industry demands (Mazurkiewicz & Poteralska, 2017).

Finally, there may be problems related to intellectual property rights. Universities and R&D centers seek to retain intellectual property for themselves, as the creation of knowledge is at the core of their work. They may, however, make the rights available for a company to use, sometimes exclusively and perpetually. On the other hand, companies also want to have intellectual property because they mistakenly understand it as the only mechanism to guarantee the exclusive and perpetual use of the developed technology (Mazurkiewicz & Poteralska, 2017).

Incentives

Finally, for authors, incentives are key when it comes to driving the science and technology ecosystem. In universities and research centers, it is necessary to incorporate metrics that recognize the contributions of researchers in the evaluation of professional and academic careers in these subjects. For companies, tax breaks present interesting incentives for investing in technological developments. It is, however, important that these incentives are easy to manage and not bureaucratic so that the industry is interested in taking advantage of them, and it becomes more attractive to get involved in developing science and technology.

To address these difficulties, it is important to recognize them and work to reduce or eliminate them. This approach should be applied both before the decision to develop the technology is made and during its development. Depending on the type of barrier, activities should be carried out individually at the level of each organization or jointly with other institutions (Mazurkiewicz & Poteralska, 2017).

Examples of Technology Transfer

The concept of technology transfer began to develop strongly in the 1980s in the United States. Years later, it was followed by Europe and, later, some Asian and Latin American countries. Additionally, the process of bringing a good or service from the laboratory to the market takes about ten years. Despite this, there are various success stories worldwide that demonstrate the relevance of this process. Some of these examples are shown below.

Gatorade

What is known as an isotonic drink is a very good example of technology transfer. In the mid-1960s, a team of researchers from the University of Florida created a drink that quickly replenished the body fluid of the players of their American football team, the Florida Gators. Consuming this liquid helped them avoid dehydration and loss of mineral salts. This incredible invention was named Gatorade in honor of the Florida Gators and scientist Robert Cade, who led the research (*Gatorade®, The World's No. 1 Sports Drink Turns 50 | PepsiCo*, 2015).

The story goes that the Florida Gators' team coach asked Dr. Robert Cade, who was not yet 40 years old, why his players could not urinate after each game. The nephrologist, who was also an excellent scientist, began to investigate and discovered that the players lost up to 8 kg of weight after each game, which lasted an average of 3 h, of which about 95% was water, due to the sweat generated by the demanding physical activity.

Robert Cade, along with Dana Shires, James Free, and Alejandro de Quesada, developed a drink that allowed rapid rehydration. It is said that the first versions tasted like disinfectant chemicals, which made the players vomit. Dr. Cade's wife suggested adding lemon juice to the drink.

In 1965, Gatorade began to be used, and the following year, appearances in the press began to make it popular. In 1967, one of Dr. Cade's students spoke with one of the vice presidents of Stokely-Van Camp, Inc., a fruit and vegetable canning company, who eventually brought the product to market.

Over the years, the Florida Gators achieved a series of victories. Massive interest in the drink was sparked in 1967 when Bud Carson (coach of Georgia Tech's Yellow Jackets) declared after his team's 27-12 loss to the Gators: "We didn't have Gatorade, that made the difference" (*Gatorade®, The World's No. 1 Sports Drink Turns 50 | PepsiCo*, 2015).

The story was not without problems. In 1970, when the use of sodium cyclamate (one of Gatorade's ingredients) was banned in food and beverages in the United States, Robert Cade and his team had to create a new formula without this compound. In 1973, a significant legal controversy arose over the intellectual property rights of the development. An agreement was eventually reached between the researchers and the university: the research team would keep 80% of the royalties, and the university would keep the remaining 20%.

Since then, Gatorade has become the quintessential hydration formula for athletes. It is currently consumed in more than 55 countries and is the official drink of various organizations.

Gatorade has also been used to manage dehydration caused by some diseases, so its contribution and impact on society is significant and corresponds to an iconic example of transfer.

Cherry Tomatoes

The cherry tomato is very ancient, and its origin is not well known. Some say it comes from Greece, others from Italy or from America in its wild form. Although it is true that the cherry tomato has existed for a long time, a group of Israeli scientists and, in parallel, an English farmer found a way to create a variety that was easy to pack, transport, and sell before it spoiled. They did this by modifying the seeds of the tomatoes to create a resistant and easy-to-grow variety. The tomatoes could reach up to 2 cm in diameter. Due to their shape and size, they were known as cherry tomatoes, as they resembled a cherry.

The story goes that during the 1970s, the owner of the English store Marks & Spencer (M&S) approached local producers and Israeli scientists to ask them to develop a small tomato that was suitable for supermarkets and had a sweeter taste. English farmer Bernard Sparkes began experimenting with a variety of cherry tomatoes called Gardener's Delight, which allowed this new variety of cherry tomato to reach the shelves of M&S in the early 1980s (Lewis, 2016).

In parallel, two Israeli scientists from the Hebrew University of Jerusalem began to produce cherry tomato seeds that grew in uniform rows and could last longer on the shelves. They improved their taste, making them sweeter and juicier, as well as making them last longer. After a while, the university mass-produced it, allowing Israeli farmers to grow it.

It is not known for certain who really created the cherry tomato. However, it is true that Sparkes managed to put it on the English market and Israel managed to mass-produce it. Today it is a success and is consumed in many countries around the world. Currently, Israel is a world leader in the tomato seed industry, and it is estimated that 10% of the global tomato seed market belongs to them.

Earthquake-Resistant Devices

February 27, 2010, was a catastrophic day in Chile. At 03:34:08 local time, one of the largest earthquakes in the country's history occurred. With a magnitude of 8.8 lasting more than 4 min, that day is remembered as 27-F. This great natural tragedy was the best test for one of the tallest towers in the city of Santiago. The

Titanium tower, with 55 floors and more than 190 m high, suffered only minor damage, without structural deterioration. It was the tallest building in the country, which passed the test of the fifth most intense earthquake in world history. This tower had an anti-seismic system based on energy dissipators.

In 1995, Professor Juan Carlos de la Llera, a civil engineer from the Pontifical Catholic University of Chile (UC), returned from his doctorate at the University of Berkeley—located on the San Andreas Fault, another highly seismic point. During his stay in the United States, he had begun to investigate technologies associated with seismic risk in different structures. He decided to promote a different line of seismic engineering in Chile than the one in place until then. It focused on two fundamental aspects which, together, separate structures from the ground: seismic isolation, with large rubber rollers; and energy dissipators, which absorb shock and dissipate vibration.

In the mid-1990s, the researcher obtained public funding to begin his research, which led to the development of more than thirty invention patents and the establishment of a company (a spin-off of the UC) in 2003.

Today, the result of years of research can be seen in action. These shock absorbers have been successfully implemented in important hospitals and buildings in Chile and have even begun to be exported to other countries, such as Peru and New Zealand.

This technology changed the life of thousands of people and made it safer. Now efforts are being made to make it more affordable to reach all corners of the world where seismic isolation is needed. This, without a doubt, is science that changes lives forever (J. de la Llera, personal communication, 2020).

Rare Diseases

The World Health Organization has defined rare diseases as all those pathologies with have a prevalence of less than 6.5 to 10 per 10,000 people. They are a major problem for patients, their families, and society in general, as they often lead to physical and emotional conditions that completely affect the quality of life of those who suffer from them. These diseases are usually not of interest to large pharmaceutical companies because the population affected by each condition is not significant compared to other diseases, making them economically unviable or of low interest.

Despite this, researchers at the Institute of Inborn Errors of Metabolism at the Pontifical Javeriana University of Colombia focus on developing therapies and diagnostics for a specific group of rare diseases: the inborn errors of metabolism.

One of the leading researchers is Dr. Carlos Javier Alméciga, whose research career has made him one of the leaders in achieving relevant innovations in this field. Carlos Javier is a pharmaceutical chemist and holds a doctorate in biological sciences, with experience in biotechnology applied to the health area, technologies related to DNA, and proteins for the therapeutic strategy for gene therapy, especially for inborn errors of metabolism.

During the final stage of his doctoral training, Dr. Alméciga participated in a research project at St. Louis University, which led him to develop an alternative therapy for Morquio A syndrome, also known as mucopolysaccharidosis type IV A. Morquio syndrome is named after the Uruguayan pediatrician Luis Morquio, who in 1929 described a family with four children affected by the same disease (Sawamoto et al., 2018). This pathology is a rare condition that causes damage to the patient's genetic information, leading to the deficient production of a protein necessary to break down complex sugar molecules called glycosaminoglycans (Khan et al., 2017). This deficiency causes physical malformations, such as short stature and skeletal alterations.

Together with the research group at St. Louis University, Dr. Alméciga developed a gene therapy-based method, which allows the genetic material used to correct the disease to be more efficiently delivered to the bones. The most interesting aspect of this technology is that it could be used to correct genetic alteration in other skeletal diseases. This technology was protected by a patent registered in the name of the universities in Colombia and the United States; a few years later, a technology-based company in the United States licensed the technology for use in the development of therapies for rare diseases.

Additionally, in 2009, the Institute of Inborn Errors of Metabolism developed and transferred to the San Ignacio University Hospital in Colombia, a series of methodologies used in the diagnosis of inborn errors of metabolism. This was achieved by adapting tests that existed at the international level to the population and technological conditions in Colombia. With the credits generated by these transfers, the Institute of Inborn Errors has created a sustainable business model, which has given rise to a virtuous circle that allows it to continue developing research and innovations with cutting-edge technologies in the region (C. Alméciga, S. Cuellar, D. Díaz, personal communication, 2020).

Water Purifier

Researcher Roland Vetter (Ph.D. in Forest Botany and chief researcher at the National Institute of Amazonian Research (INPA) in Brazil) developed a technology called "solar dryer," which is a device for drying wood using solar energy. The idea emerged to meet the needs of various communities that did not have electricity and sought to replace conventional wood dryers that consumed much energy. In 2005, while visiting an indigenous community in Brazil with the aim of presenting the technology, he was surprised to see the community's need to improve water quality; many children and older people were experiencing health problems such as diarrhea and cholera due to water pollution from streams and rivers used by the community. From this need, the researcher and his team sought alternatives to disinfect water, using the technology devised for the solar dryer. The result was the design of a solar-powered water purifier.

In 2009, this technology was completed and named *Água Box*, protected by a utility model patent, owned by INPA and inventor Dr. Roland Vetter. The water purifier is a 13-kg metal box that operates with a water pump and two 90-W solar panels that pump water from the river to a water reservoir. The contaminated water passes through a filter and then through a stainless-steel tube with a type C ultraviolet lamp, which eliminates the DNA (Deoxyribonucleic Acid) of microorganisms that cause intestinal diseases. After these processes, the water is drinkable, germ-free, and ready for consumption. The lamp and battery last about 10,000 h, or three to four years. The purifier removes 99.5% of bacteria, fungi, and fecal coliforms from river water using a type C ultraviolet lamp, the most dangerous rays of ultraviolet radiation. Connected to a solar panel and the community's water tank, the purifier filters 400 L per h, or 5000 L per day—enough to provide clean water for drinking and cooking for 300 people. The system's price is less than $1000 (US), including the solar panel and input filter.

The Água Box was implemented in riverside communities in Brazil's Amazon, in the Deni and Yanomami indigenous villages, and in the Nampula region of Mozambique in Africa.

Between 2012 and 2017, INPA licensed the technology to the Manaus company, Q'Luz EcoEnergia, and it was renamed *Ecolágua*, becoming the institution's first technology to generate royalties.

In 2015, INPA partnered with the Brazilian Army to develop a project called *Hombre Água*, which consisted of creating a more compact model of the water purifier that fits in a backpack. The adapted version weighs approximately eight kilograms and has the capacity to purify up to 300 L per h.

In 2017, the patent was granted, and in 2020, the technology was licensed to a new company.

Human Papillomavirus Vaccine (HPV)

Human papillomavirus is a genital virus transmitted by direct skin contact. There are many types of HPV, and most of them do not cause symptoms; however, some can cause cervical cancer. The American Cancer Society estimated that in the United States, in 2022, more than 14,000 new cases of cervical cancer would be diagnosed, almost all due to HPV, and more than 4000 women would die from this cause.

In 1976, Spanish gynecologist Enrique Aguirre Cabaña discovered the relationship between HPV and cervical cancer. In the following decade, a group of researchers from the universities of Rochester and Georgetown, and the National Cancer Institute of the U.S., and the University of Queensland in Australia, began research for the development of a vaccine against this virus. In 1991, researchers Ian Fraser and Jian Zhou from the Australian University developed non-infectious virus-like particles that activated the immune system.

The scientific development was not without problems. There is much controversy surrounding who the creators of this invention were. Although that is true, there is consensus that the development was carried out jointly among the four organizations; the patent in the United States recognized the four institutions as applicants. After ten years, the dominant patent was granted, from which the HPV vaccine patent was based. However, Rochester, NIH, and Queensland have patents in other countries of the world (McNeil, 2006).

In 1994, the technology transfer office of the University of Queensland, UniQuest, licensed the technology to the Australian biotechnology company CSL, granting the license for the use of this technology to CSL (the largest Australian biotechnology company). The technology was later acquired by the company Merck & Co. Inc.

The vaccine was incorporated into the official vaccination schedule in several countries worldwide and today prevents thousands of deaths each year.

Ideation Process

Technology Push Versus *Demand Pull*

According to Brem and Voigt, (2009), ideas that can be transformed into science-based innovation can come from two sources. One of them is a market need, known as market-pull or demand-pull, which means that the company dictates what is required. The other source is scientific curiosity, known as technology-push, according to which researchers decide what to investigate based on what they believe is necessary.

Therefore, market pull, or demand pull is the mechanism from which ideas that are transformed into technological developments arise and whose origin begins with a need of the company or the market in general. Once identified, R&D organizations, such as universities and R&D centers, begin scientific development to create a solution for this need (which can be specific or general) and solve a specific problem of a particular company or a general problem of industry.

On the other hand, those known as technology push are those that arise from intellectual processes within R&D organizations without necessarily having clarity about the market need; that is, they arise from dissatisfaction or scientific curiosity of the researcher, and then the goal is to commercialize the new good or service (Brem & Voigt, 2009).

Reality, however, is not so black and white; generally, these ideas arise from a combination of both mechanisms. Gone are the days of "Eureka!"—the famous expression of Archimedes, the Greek mathematician who uttered it after discovering that the volume of water that rises is equal to the volume of the submerged body, which led him to solve the problem of measuring the volume of irregular bodies. Since then, this phrase has been used to celebrate a significant discovery or finding. When talking about technological development, researchers relate to their environment either through scientific congresses and conferences or in dialogue with the industry, which provides them with relevant information for their research. Therefore, it Is uncommon for breakthroughs to occur in the isolation of the laboratory and elicit the excitement equivalent to Archimedes' cry of "Eureka!".

On the other hand, companies do not always have clarity about their problems, and depending on the country, they do not always have the vision, the will, the financing, or the incentives to approach R&D organizations and request specific developments. Therefore, this approach usually occurs between both parties and with a combination of "Eureka!" moments and industry requirements.

Undoubtedly, both mechanisms have certain deficiencies and limitations. In the case of technology push, there is the risk of starting research with what is easiest for the researcher—developing solutions for needs of little interest to society or the market and focusing on the development of a solution for a problem that does not exist. On the other hand, the demand-pull can focus on those limited needs that are easily identified but have a lower potential or impact.

According to Ameka (2013), one of the aspects that strongly influences the development of science-based innovation is the proactivity of the business sector in linking with R&D organizations, i.e., expressing their development needs and, in turn, knowing what risks they are willing to take in this race for the development of innovations. In this sense, the success of a new good or service that originates from an R&D process depends on the harmony and direction that exists between the demand pull and the technology push. To achieve this, networking (the ability to relate or link) is recognized as the essential key to success.

Bootleg Research

Another relevant concept to consider in the emergence of ideas is so-called bootleg research or "clandestine research." This mainly affects companies, and refers to a way of developing an idea within an organization without its formal authorization or consent, or against what the organization establishes.

According to Criscuolo et al. (2014) in their research on bootlegging, organizations often have to deal with finding a balance between allowing their collaborators, specifically those dedicated to R&D, enough flexibility and autonomy to explore the *new* and *unusual*, and keeping them controlled to ensure that their innovative efforts are aligned with the company's strategies, objectives, and priorities. In cases where there are many restrictions, individuals may seek to increase their autonomy by taking part their work *underground*. These clandestine activities are described as bootlegging.

According to Augsdorfer (2005), bootlegging is defined as the process by which individuals take the initiative to work on ideas that do not have formal backing from the organization and are often hidden from the view of top management but are carried out with the aim of producing innovations that benefit the same company. In this context, bootlegging is a vehicle that allows researchers to achieve greater autonomy in defining their research agendas and objectives in environments where this is not formally granted (Criscuolo et al., 2014).

The author continues to argue that in the case of companies and R&D organizations with specific development focuses, formal rules and procedures can limit the flexibility and creativity required to explore new and valuable opportunities and limit the scope of experimentation. However, the desire to ensure autonomy in the pursuit of new ideas and the development of innovations must be balanced with a level of responsibility for the roles assigned to the collaborators involved.

Two types of autonomy are particularly relevant: operational autonomy and strategic autonomy. Because the typical R&D organization is characterized by highly skilled people working on complex and difficult projects, it is common for researchers to have high levels of operational autonomy in their work. However, only some organizations formally grant a degree of strategic autonomy that gives research teams the opportunity to define their own research agendas and objectives and establish the overall direction of their projects (Criscuolo et al., 2014). While this is mostly true in universities, which generally have the principle of research freedom, there are also some companies, like Google, where developers can dedicate 20% of their time to researching things outside their research agenda.

Unlike these particular cases, as a result of the restrictions imposed by organizations, researchers may resort to informal methods to obtain greater strategic autonomy, secretly participating in unscheduled innovation efforts (Criscuolo et al., 2014). In this sense, bootlegging emerges as a continuous incremental learning process of trial and error in R&D, and its uncertainty is no better or worse than that of *authorized* R&D (Augsdorfer, 2005). This practice often arises because the idea is difficult to justify and quite risky in the eyes of the organization.

The value of bootleg research for a company can be tangible and intangible. While the tangible results of research have clear economic benefits in terms of increased profits for the company, the intangible benefits consist of, on the one hand, increasing the technological knowledge base through the building of competencies and, on the other hand, keeping research teams motivated (Augsdorfer, 2005).

The study points out that in the initial stage of innovative activity, there is very little difference between bootlegging projects and authorized projects. The first steps of research generally consist of data collection, either in the form of small experiments or literature reviews. The difference is that bootlegging represents a low level of uncertainty in terms of resources; the research incurs only minor additional costs in terms of time consumption and the cost of tools, materials, and equipment, as these are often already available in the laboratory.

In summary, managers who fear losing control can be assured that there is nothing to fear. Ideas do not develop by accident or in isolation. New ideas are somehow related to what researchers already know (Augsdorfer, 2005).

There are multiple examples that demonstrate that this practice has meant significant benefits for the organization, and of course, many account for it being a resounding failure and a loss of important resources. However, it is always better and more beneficial to develop authorized R&D supported by the organization.

Historical Context and Policies Supporting Technology Transfer

The United States

Although the history of technology transfer in the United States dates back a long time, the fact is that in the 1980s, a turning point was reached that resulted in a strong momentum. In the following years, history showed that this momentum occurred not only in the United States but worldwide. In that decade, Congress passed the historic legislation governing university technology transfer called the Patent and Trademark Law Amendments Act, better known as the Bayh-Dole Act, which regulates intellectual property arising from government-funded research. This law was promoted by Senators Birch Bayh of Indiana and Bob Dole of Kansas. Birch Bayh assumed office as a Democratic senator at the age of 35, won two re-elections, and lost the third, which led him to hold office between 1963 and 1981. Bob Dole, a Republican senator, held office between 1969 and 1996.

This law drastically changed the incentives for universities and companies to engage in university technology transfer and simplified the commercialization process by establishing a uniform patent policy and eliminating many restrictions on licensing government-funded technologies. More importantly, it allowed universities to own patents arising from federal research grants, which until that date was not permitted. This law is considered to have not only transformed the science-based innovation system in the United States but also laid the legal foundation that allowed recipients of government funds for R&D activities the right to patent inventions and grant licenses, strongly encouraging university-industry linkages and facilitating the market entry of goods and services based on research results (Link et al., 2015).

With this law, the authors say, universities were encouraged to start creating units responsible for responding to the need to manage the transfer process formally, with clear incentives and policies, generating long-term links with industry and the public sector. Most research universities have technology transfer offices (also known as transfer and licensing offices), and universities are increasingly becoming engines of economic growth and local and regional development through the commercialization of their intellectual property.

Although the effects of the Bayh-Dole Act on the increase in patents are far from definitive and conclusive, universities and other public research organizations are increasingly protecting their inventions, ranging from genetic developments to *software* programs, with the expectation of generating additional funds for research, and financing part of university activity. For example, before 1981 universities applied for fewer than 250 patents each year; just over a decade later, they applied for nearly 1600 patents each year (Henderson et al., 1998; Shane, 2004). Another relevant piece of data related to patents is that between 1993 and

2000, approximately 20,000 patents were granted to U.S. universities. During that period, some of these university patents generated millions of dollars in licensing revenue and stimulated the creation of more than 3000 new companies.

According to data from the association of university technology managers (AUTM), between 1996 and 2017 technology transfer in the United States has supported the creation of nearly six million new jobs and more than 13,000 science-based companies; it has contributed $865 billion to the country's gross domestic product (Atkins, 2009).

AUTM conducts an annual survey (*AUTM Licensing Survey*) among its member organizations. The results for 2020 showed that total research expenditures increased to $83.1 billion—an increase of more than 22% in the last five years. Patents increased by more than 15%, reaching 8706 that year. Additionally, 1117 new companies were created in 2020 (AUTM, 2020).

Europe

In Europe, there are significant differences in terms of university intellectual property management, although some convergence can be found, according to the study conducted by Siegel (2015). For example, the United Kingdom has increasingly strengthened universities' rights to intellectual property generated by their academics by passing regulations that consider the distribution of associated royalties among those who participated in the creation of research results and the university areas that participated in the development.

The study states that Germany and Belgium adopted Bayh-Dole type legislation in the late 1990s, while in France, this type of regulation had already existed for some time. In Italy, public sector researchers receive property rights for their developments; however, in most cases, they enter into a formal contract with individual universities to grant intellectual property rights to the university. In Sweden and Finland, discussions are underway to change their policies and govern them using a Bayh-Dole type model rather than the inventor ownership model.

On the other hand, most European countries have changed their legislation to make it possible and more attractive for researchers and academics to take equity stakes in companies created to develop and commercially exploit their results, and to be able to receive royalties or income from commercialization. In France, for example, before 1999 it was illegal for an academic to take an equity stake in a company (Link et al., 2015).

The study adds that, unlike what happens in the United States, the level of commercial activity in European universities is relatively low compared to the high levels of scientific performance and investment in research they have. This phenomenon is known as the "European Academic Paradox." To a large extent, this awareness reflected the recognition by governments that, in some cases, placing the results of State-funded research in the public domain is not enough to generate social and economic benefits from research. Thus, since the early 2000s, many

European countries, such as Austria, Denmark, Germany, and Norway, have eliminated the so-called professor's privilege, which granted academics the right to own patents. The right to intellectual property has now been transferred to universities, while academic inventors receive a share of the income via royalties, according to the Bayh-Dole model.

Latin America

The case of Latin America is more recent and diverse, as each country has experienced different realities. For example, in Argentina, although it is true that the first activities emerged in the 1980s, the starting point of institutionalization is considered to be the creation in the 1990s of the State-level technological linkage unit, which was a pioneer in the Latin American region. More recently, and due to the need to design and execute a comprehensive policy for promoting technological linkage and transfer, the technological linkage offices were created in 2013 within the Argentine scientific and technological centers, whose purpose is to promote the commercial exploration of research results.

In 1981, in Brazil, the Nucleos de Inovação Tecnológica (NITs) were designed with the aim of improving links between academia and industry; despite State efforts, the initiative was suspended at the end of the 1980s (Santos, 2016). In the same decade, the activities of the Support Program for Scientific and Technological Development (PADCT) began; linked to the country's Ministry of Science, Technology, and Innovation, its areas of action included industrial property. In 1998, a debate began on the importance of intellectual property for Brazil's economic development, particularly in relation to economic internationalization. In that decade, the lack of national legislation that considered and guided the actions of this developing area was evident: the Centers for Technological Innovation (TIC) received different names and diverged in actions that were completely unfocused. This situation was crucial to the creation of Brazil's National Innovation Law in 2004, inspired by the Bayh-Dole Act (Chais et al., 2017), whose objective is to manage innovation policy, monitor development processes from research to innovation, and promote partnerships between universities and companies.

In the case of Chile, the history is even more recent. Although it is true that the country has an intellectual property law dating back to the 1970s, it was only at the beginning of this century that some universities began to create units responsible for managing intellectual property and technology transfer.

In that country, efforts to promote technology transfer were generated by the state; thus, at the beginning of the 2000s, the first attempts were made to create institutions and begin the process of training people to manage transfer offices in the following years. At the beginning of the second decade, the government agency responsible for promoting innovation and entrepreneurship called the Corporation for the Promotion of Production (CORFO), dependent on the Ministry of Economy, launched funding for the creation and strengthening of marketing and technology transfer capacities within institutions. In this context, it called for the first competition to strengthen the Technology Transfer and Licensing Offices of national universities and scientific-technological centers. Initially, more than twenty institutions were supported, which developed new intellectual property policies, procedures, and regulations, expanded their contact networks, and built internal capacities.

Since 2011 significant efforts have been made to create technological management capacities in universities and scientific-technological centers, with important institutional advances in intellectual property management and formalization of technology transfer processes; however, only recently has there been an increase in the number of patents and licenses. The results in technological contracts and the creation of technology-based companies are still incipient, as these are processes that take time to mature.

In the region, countries such as Mexico, Colombia, and Peru have been making significant efforts in the field of intellectual property and technology transfer, which, as in other Latin American countries, have been strongly inspired by the experience of the United States.

References

Ameka, I. (2013). Technology push vs. market pull in technology university innovation commercialization case study: ITB. *Information Management and Business Review, 5*(7), 337–341. https://doi.org/10.22610/imbr.v5i7.1060

Atkins, M. E. (2009). Driving the innovation economy. *Baseline, 94*, 11. http://search.ebscohost.com/login.aspx?direct=true&db=bsu&AN=37323636&site=ehost-live

Augsdorfer, P. (2005). Bootlegging and path dependency. *Research Policy, 34*(1), 1–11. https://doi.org/10.1016/j.respol.2004.09.010

AUTM. (2020). *Licensing activity survey* (pp. 3–24). https://autm.net/AUTM/media/SurveyReportsPDF/FY20-US-Licensing-Survey-FNL.pdf

Biggar Economics. (2017). *The economic contribution of the institutions of the LERU universities* (Vol. 44).

Brem, A., & Voigt, K.-I. (2009). Integration of market pull and technology push in the corporate front end and innovation management—Insights from the German software industry. *Technovation, 29*(5), 351–367. https://doi.org/10.1016/j.technovation.2008.06.003

Chais, C., Ganzer, P. P., & Olea, P. M. (2017). Technology transfer between universities and companies: Two cases of Brazilian universities. *RAI Revista de Administração e Inovação*, 1–13. https://doi.org/10.1016/j.rai.2017.07.003

Criscuolo, P., Salter, A., & Ter Wal, A. L. J. (2014). Going underground: Bootlegging and individual innovative performance. *Organization Science, 25*(5), 1287–1305. https://doi.org/10.1287/orsc.2013.0856

Fraser, J. (2008). Communicating the full value of academic technology transfer: Some lessons learned. *The Licensing Journal, 28*(1), 1–10. http://www.technologytransfertactics.com/content/audioc/6252008/JAF-Licensing-Journal-article.pdf

Gatorade®, The World's No. 1 Sports Drink, Celebrates 50 Years | PepsiCo. (2015). https://www.pepsico.cl/es-CL/live/pressrelease/gatorade-la-bebida-deportiva-n-1-del-mundo-cumple-50-años

Henderson, R., Jaffe, A. B., & Trajtenberg, M. (1998). Universities as a source of commercial technology: A detailed analysis of university Patenting, 1965–1988. *Review of Economics and Statistics, 80*(1), 119–127. https://doi.org/10.1162/003465398557221

Khan, S., Alméciga-Díaz, C. J., Sawamoto, K., Mackenzie, W. G., Theroux, M. C., Pizarro, C., Mason, R. W., Orii, T., & Tomatsu, S. (2017). Mucopolysaccharidosis IVA and glycosaminoglycans. *Molecular Genetics and Metabolism, 120*(1–2), 78–95. https://doi.org/10.1016/j.ymgme.2016.11.007

Lewis, D. (2016). The contentious history of the cherry Tomato | Smart news | Smithsonian magazine. *Smithsonian*. https://www.smithsonianmag.com/smart-news/contentious-history-cherry-tomato-180959420/

Link, A. N., Siegel, D. S., & Wright, M. (2015). *The Chicago handbook of university technology transfer and academic entrepreneurship*. University of Chicago Press. https://doi.org/10.7208/chicago/9780226178486.001.0001

Mazurkiewicz, A., & Poteralska, B. (2017). Technology transfer barriers and challenges faced by R&D organisations. *Procedia Engineering, 182*, 457–465. https://doi.org/10.1016/j.proeng.2017.03.134

McNeil, C. (2006). Who invented the VLP cervical cancer vaccines? *Journal of the National Cancer Institute, 98*(7), 433. https://doi.org/10.1093/JNCI/DJJ144

Pigna, F., & Pacheco, M. (n.d.). *Bernardo Houssay, father of physiology and great promoter of research in our country*. The Historian. Retrieved December 19, 2020, from https://www.elhistoriador.com.ar/bernardo-houssay-padre-de-la-fisiologia-y-gran-promotor-de-la-investigacion-en-nuestro-pais/

Santos, M. E. R. (2016). The University as an instrument of development and technology transfer. In *Integrity in research and intellectual property in the university* (pp. 81–108). EDIPUCRS.

Sawamoto, K., Alméciga-Díaz, C., W, M., Mason, R., Orii, T., & Tomatsu, S. (2018). In S. Tomatsu, C. Lavery, R. Giugliani, P. Harmatz, M. Scarpa, G. Węgrzyn, & T. Orii (Eds.), *Mucopolysaccharidosis type IVA: Clinical features, biochemistry, diagnosis, genetics, and treatment* (Vol. I, pp. 235–272). Nova Publishers (Chapter 12).

Shane, S. (2004). Encouraging university entrepreneurship? The effect of the Bayh-Dole act on university patenting in the United States. *Journal of Business Venturing, 19*(1), 127–151. https://doi.org/10.1016/S0883-9026(02)00114-3

Commercial Validation

2

Abstract

Focused on "commercial validation" in identifying market opportunities, this chapter highlights its crucial role in the success of an invention or research outcome in the market. Commercial validation becomes a determining factor in ensuring the triumph of an innovative idea. Likewise, the importance of identifying the various applications of technological development that can generate a significant impact on different industries and capture the attention of potential customers is explored. In addition, the need to precisely define the problem to be solved and analyze its scope in society is emphasized. The "value proposition" emerges as a fundamental element to differentiate a product and meet the needs of customers. In this sense, it is essential to focus on the functional and emotional benefits offered by the product, formulate it clearly, and direct it towards a specific group of customers. Building a solid value proposition involves a deep understanding of how customers' problems are alleviated and the ability to present a unique and attractive solution.

Keywords

Commercial validation • Market opportunity • Value proposition • Minimum viable product

Relevant Aspects of Commercial Validation

One of the fundamental aspects to increase the chances of an invention or research result reaching the market is to validate it commercially; that is, to verify if an appropriate solution is being developed for a real problem. To this end, there are different mechanisms, depending on the type of technology or the state of development progress, the most important thing being that this process begins at the earliest possible stage.

A. Ossa, *From Research to Market*, https://doi.org/10.1007/978-3-031-71340-8_2

The first thing to validate is the *opportunity*. This concept is also known as *problem*. That is, someone, whether a natural or legal person, has a specific need that requires a solution. To validate this opportunity, it is important to talk to the entire chain of relevant actors, such as users, customers (whether intermediate or final), suppliers, or competitors, and collect as much information as possible in order to understand and make a thorough diagnosis to focus or guide the solution.

There are multiple ways to carry out these validations. One of them is the so-called expert validation, which consists of conducting interviews with people knowledgeable in the technical area but fundamentally in the commercial area. This can be done by gathering a group of people or having meetings or telephone conversations with each one individually. In general, it is preferable to contact users who are in intermediate processes, that is, those who facilitate the good or service reaching the end user. This recommendation is important, as it is very complex and costly to interview end users at such early stages due to the volume of responses needed to be representative or because, given the early stage of development, it may be better to obtain more information about the process or value chain to reach the end user. The important thing about this process is to listen to the needs and, based on that, guide or redirect the potential solution.

According to what was stated by García et al. (2014), in a validation process, it is very important to really understand how users suffer or feel a problem. This is supported by MIT academic Otto Scharmer in his Theory U: "When you listen to your potential customer, open your mind and do not draw hasty conclusions, as you could judge reality with the same approach you have always done" (García et al., 2014). Scharmer calls this the voice of judgment. The second voice that appears in this process is that of cynicism, "which makes you protect yourself from the observed reality so that the problems do not affect you, and the third is that of fear, which blocks and limits your possibilities of creation. Now that you know these three voices exist, try to omit them when you listen to your potential customers" (García et al., 2014).

The same author states that it is very important to be aware that problems are difficult to identify due to two factors: their nature, that is, their complexity, uncertainty, or ambiguity, or the way they are seen or approached, either by biases, prejudices, history, ideology, preferences, interests, or emotions, among other factors. Finally, before starting validation with other actors, it is very important to gather information from secondary sources. This means obtaining already organized and analyzed information that was obtained from primary or first-hand sources, such as market studies, reports, research, or others (whether paid or free), in order to be prepared and thus obtain better results from the meeting with expert panels.

The second fundamental aspect is to validate the proposed solution. Once the problem has been identified and more deeply understood and sufficient information has been gathered, the possible solution must be validated. In this sense, and continuing with the arguments previously presented, it is advisable to continue validating with intermediate users. At this stage, it is important to consider technical characteristics, regulatory aspects, and key phases.

When analyzing the possible solution, one of the main challenges researchers face is how to move from research to commercialization. As Sethi (2016) states in the book *From Science to Startup*, researchers do it for the sake of research and rarely focus on accelerating something towards the market. Their main motivation is often to publish their articles in scientific journals and obtain peer reviews rather than potential users.

It is very common that at such early stages, not all questions may have answers. But it is highly relevant to ask them in order to have clarity on how much is known and how long it might take to achieve a marketable product. This text raises some fundamental questions to minimize the risks of any unforeseen events that may hinder or delay commercialization, such as: What steps are required to start working on a pilot test? Do any of the steps require fundamental research, or is it enough to develop manufacturing capacity? Is it replicable? Are there regulations governing manufacturing? (Sethi, 2016).

Therefore, it is important to understand that the commercial validation process is iterative. It fluctuates between understanding the problem, validating with those who experience it, moving toward a feasible solution (both technically and commercially), and returning once again to understanding the problem—until the most appropriate solution is reached.

Market Opportunity

The "market opportunity" is not based on technological development but rather on the applications of such development, which can impact different industries and different customers. In other words, a market opportunity is any combination of application and customers; the link between them is called a *need* (Gruber & Tal, 2017).

The authors argue that a diverse set of market opportunities is an asset in itself, as it increases the chances of detecting the most promising option. Identifying a set of potential market opportunities allows one to choose the most attractive option and, thus, not waste time with less relevant opportunities. This process is not simple, especially in those developments that have their origin in scientific curiosity rather than market demand.

Identifying possible market opportunities is a discovery process that requires questions, networks, and even experimentation to understand the different needs of customers and how to address them. To evaluate them, it is necessary to analyze their potential to impact the market and the risks involved (Gruber & Tal, 2017).

The potential of a market opportunity—the authors continue—is made up of three main factors: the first is the compelling reason to buy, also known as the value proposition, which will be discussed later; the second is the market size, that is, how big is the segment to which we want to target the product; finally, there is economic feasibility, that is, the probability that the product can succeed. This allows one to understand how big the opportunity is and how much value can be created.

On the other hand, to analyze risk, it is necessary to understand how challenging it is to pursue the opportunity, what obstacles lie ahead, the main challenges, and how difficult it is to overcome those challenges and conquer the opportunity (Gruber & Tal, 2017).

According to the authors, to determine an adequate market opportunity, it is also necessary to know the customers' environment, understand the value chain and competitors in the business environment, and understand the key implementation milestones that still need to be developed to create and deliver the product. Since there may be multiple opportunities, market selection is crucial for creating value. It is worth mentioning how difficult this phase can be, given the uncertainty that exists and the constant changes in the environment.

Therefore, the authors conclude that a strategy needs a proper balance between focusing on the detected opportunity for technological development and the flexibility to remain consciously open to other market options that might be more suitable. Undoubtedly, the challenge is in choosing the market opportunity, but it is also necessary to be flexible and available to explore other options.

The key factor when analyzing market opportunity is being aware of the existing risks, which constitute the first step to face them.

Problem and Solution

Researchers spend a lot of time finding the ideal solution. However, when they find it, it often turns out to be a perfect solution for a problem that does not exist.

To properly define the problem, it is necessary to analyze it in depth. It needs to be determined whether or not it is really a problem, understood for whom it exists, know if it is associated with other problems, and delimited in scope in order to address it.

In the process, it is advisable to discuss it with industry actors, and then strictly analyze their opinions in order to structure the problem in the most realistic way possible. When talking to different people, it is important to identify their industry role in order to enrich the analysis with opinions from more actors with different perspectives.

Another important factor that helps to properly define the problem is to delimit it from both a technical and commercial point of view since circumscribing allows for a more suitable solution. If the problem is very broad and there are multiple factors influencing its definition, it is much more difficult to address it and develop an appropriate solution.

It also helps to define the problem well if it knows whether there are other problems associated with it and, if so, what the order of priority is. This way, it can begin to gauge its magnitude and importance.

Finally, it helps define key actors. For example, it should know who the people with whom the problem should be discussed and clearly identify who the users and who are the customers (in case they are different), as well as determining their characteristics, and defining the intermediaries.

Now, with the problem well-identified, the possible solution must be analyzed. Since there are multiple ways to solve the same challenge, it is important to verify that the proposed solution is the most suitable.

It is essential to start by evaluating the most general functionalities or characteristics of technological developments and then try to understand what other applications with these characteristics are feasible for different types of customers. The characteristics of the technology correspond to the set of resources and capabilities that the scientific development currently has, whether they are currently being developed or are planned to be developed.

To do this, it must be very clear what the solution consists of (from a technical point of view) and its state of development. This allows for the analysis of the solution's flexibility to adapt to any changes to the problem or even solve other problems that may be associated with it.

It is also necessary to analyze how the problem is currently solved, and essential to compare the current solution with the proposed solution, in order to understand the advantages of the proposed solution.

Finally, it is crucial to know the unique qualities of the solution—the differentiating attributes compared to the existing ones. This allows the solution to be assessed and generates an appropriate value proposition for customers.

Value Proposition

The value proposition is the set of both functional and emotional benefits of a product that makes it unique and different from what exists. It represents the promise made to the customer, and is the concrete expression of the benefits they receive. Here, no reference is made to the product, but instead to the solution it provides to the customers' current problems (Osterwalder et al., 2014).

This must be centered on the needs and desires of the customers, so it makes no sense to develop generic value propositions aimed at unknown customer groups. Since customers are increasingly demanding if the value proposition is not clearly formulated and well-identified, they are not interested; they understand that the offer is not directed at them.

The value proposition is a novel development that meets existing needs in the market and society, and provides benefits to customers. To build a good value proposition, it is important to address essential things, such as knowing what is being offered, understanding how the customers' problem is being alleviated' what benefits the good or service brings to them, and why they should buy this product and not another. The focus should be on the benefits that matter most to customers and for which many people are willing to pay. In this sense, one should not try to solve all problems with a single product; it is important to target a few problems and benefits and solve them extremely well.

The proposal starts by moving away from formulating the strategy based on what is being manufactured or what researchers know how to do. The key is that the value proposition must be conceived from the customer's perception of value. The starting point should be to perfectly understand the criteria a customer uses to evaluate the current offer.

A good exercise to analyze if the value proposition is well-defined is to state it in the following way and see if it makes sense to the customer or not: The [*detail the good or service*] helps [*specify the customer segment*], who want [*name the activities to be performed*] to [*use verbs like reduce, build, or others*], unlike [*competitor's value proposition*].

Minimum Viable Product

A very relevant concept within the commercial validation process is the "Minimum Viable Product," which refers to a product prototype that has the minimum functionalities to present it to customers and verify their interest.

Seen another way, it is a development with enough features to satisfy initial customers and provide feedback for future development, i.e., a product that someone would be willing to pay for or adopt to continue its development.

Achieving a minimum viable product is not always easy when developing technology; research teams generally seek perfection in development and are not able to validate with a third party something that is not yet finished. For this reason, it becomes very important to understand how much development is needed for the research result to become a minimum viable product that can be validated in a real environment.

It is essential to discern and agree on what features the research result must have in order to achieve a minimum viable product state. Part of the early validation process involves discussions with key industry stakeholders in order to get the feedback required to make those determinations.

Once the current state is determined, and knowing how much is left to obtain a minimum viable product, it is necessary to analyze what is needed to reach that point in terms of time, financial resources, or technical capabilities.

Technical and Commercial Analysis

To develop a transfer strategy, it is necessary to evaluate whether the research result meets the necessary conditions to reach the market in the form of a good or service. This requires a feasibility analysis, which reviews its main aspects in order to determine whether or not the research result has the possibility of being transferred.

The technical and commercial feasibility analysis determines the state in which a research result is at a given moment. That snapshot, in turn, make it possible to design a plan aimed at working on each of those interrelated aspects, in order to bring that result to a stage closer to a product.

The feasibility analysis is necessary because the academic merit of a research result does not imply or guarantee that it will be transferred. For example, it may be that technically it is a great solution but there is not a large enough market; or that, being a technically very attractive solution, there are regulatory barriers that prevent its implementation; or that it is very costly to turn the research into a product; among many other reasons.

In this sense, effectively transferring a research result depends on a large number of aspects, which cannot always be controlled by researchers. However, it is possible to increase the chances of a successful transfer by focusing efforts on knowing and adequately addressing such aspects.

There are five fundamental aspects to consider in the feasibility analysis: the state of development, financing, regulatory aspects, intellectual property, and the market.

Development Status

It is essential to understand the development status of the invention and to do so; the first step is to understand exactly what the result of the existing research is. For example, suppose the development is a wildfire predictor. In that case, it is important to define what the technological development consists of, whether it is *software* or an algorithm, or if it is an interface –or a combination of all of them. The important thing is to clarify what is particular and what is essential about the developed technology.

Once understood what is *there*, it must be determined what is *missing* in terms of development for the research result to become a good or service that someone would be willing to pay for. Continuing with the previous example, it would be necessary to evaluate if there is an appropriate user interface or, if not, how far it is from being a user-friendly interface.

Another element to analyze is the level of development and the appropriate state to enter the market. A useful scale to determine the development status of a technology is the Technology Readiness Level (TRL). The TRL is an international scale used to measure the maturity of technological projects. It includes ten levels ranging from the idea to the finished product, ready for marketing to the end user (Ibañez, 2014). TRL0 is the idea; TRL1 is the postulated and observed principles; TRL2 is the formulation of concepts and hypotheses; TRL3 is the first laboratory tests completed; TRL4, the prototype is at a small scale; TRL5, the prototype is at a large scale; TRL6, the prototype is field-tested; TRL7, the system works and is operational; TRL8 is the first commercial version; TRL9, the commercial application is complete.

Once the current status is determined, a plan can be made that includes the necessary resources in financial, material, and personnel terms to reach the necessary development level to transfer the technology.

Intellectual Property

It is important to determine whether or not the research result can be protected by intellectual property rights, to guarantee its eventual use and exploitation, and to establish the best protection strategy.

There are multiple intellectual protection routes. For example, patents, which are a powerful mechanism to prevent third parties from manufacturing, exploiting, or selling the invention for a certain period, which may depend on the type of invention; trademarks, which prevent other companies from selling a product, or offering the same service under the same name; and, copyright, which allows the creator to control the reproduction, distribution, display, or performance of their work, and is granted automatically without the need for official registration.

Since intellectual property rights are not mutually exclusive, it is recommended to design a comprehensive protection strategy in order to generate more robust protection that allows improving the competitive position in the market.

Having protection through intellectual property rights allows establishing conditions for the use or exploitation of the research result. It is for this reason that the robustness of the protection strategy and the vulnerability or strength of the market position must be analyzed.

Another important aspect to analyze at this stage is Freedom to Operate (FTO). This refers to whether or not it is commercially safe to manufacture or sell the technology in the country where one wishes to do so without infringing on the rights of others. If there are valid registered rights, there is a risk of not being able to carry out the activity, or being sued for possible infringement, and even having to pay damages.

Market Analysis

The first thing to do when conducting a market analysis is to understand the need; evaluating the problem facilitates understanding of the market. The strategies is different if a problem is relevant to, for example, small or large agricultural producers or if they are in a specific geographic region or scattered throughout the country.

It is also necessary to analyze the magnitude and economic impact of the problem to determine if it is a global problem; if the market is large, and how large it is; whether or not it implies a real economic impact; if the problem affects multiple countries (which could translate into a considerable economic opportunity), unlike a problem with strictly local impact. If the problem affects one country, it has a small impact, but if it affects all countries with similar characteristics, then there is a larger economic opportunity.

As previously mentioned, after validating the problem, it is critical to analyze whether the solution is perceived by the involved actors as adequate, and whether it benefits them or not.

For example, if a highly relevant problem requiring resolution has been identified, that does not necessarily indicate that the proposed technology is perceived as adequate. Perhaps a new construction material is being developed, but the assembly system is very costly, or its implementation contradicts existing regulations (such as height efficiency), or it does not comply with the country's seismic standards, or simply the market does not *believe* in the technology because it is not prepared to adopt it.

An analysis and comparison with existing solutions, both qualitative and quantitative, should also be conducted. Generally, each research team considers their research result to be unique, and this can indeed be the case in most instances. However, this does not invalidate the existence of other solutions for the same problem in the market, or that these solutions are in development. In this sense, it is advisable to analyze not only competitors but also substitute technologies. Regardless of the type of solution that exists in the market, each one is a precedent to consider and compare with the proposed technology in order to determine the reasons why that particular development could be better.

Likewise, when comparing the proposed solution with existing ones, it is not enough to simply determine if it is better, more economical, or reduces certain times. It is important that the comparison is not only qualitative but also quantitative and very precise.

In addition, comparisons must be related to the value proposition, which in turn must be built from the analysis of the customer segments targeted by the solution, the type of relationship expected to be established and maintained with them, etc.

The first market that the product will target must be clearly defined. It sounds easier than it is, as there can be a series of attractive, approachable markets. The most difficult decision may be whether to opt for high-value or high-volume markets.

Regulatory Aspects

There are regulatory aspects that condition certain technologies from reaching the market, so it is necessary to analyze if there are any regulations, certifications, or quality standards that must be met.

When conceiving a research project, it is important to consider these regulations from the beginning of the research development; it may be that a technology is not attractive to the industry, or even unfeasible from a practical standpoint, if these factors are not considered.

There are not always regulatory aspects or quality standards that must be met. Even so, in some areas such as health, agrochemicals, energy, and food, the existence of regulations is much more common.

To obtain approval for the commercialization of new pharmaceutical drugs, medical devices, diagnostic kits, plant varieties, pesticides, cosmetics, or food, the regulatory entities of each country establish standards that must be met. These could vary from country to country and also over time.

Thus, a new technology with a significant impact in an area but not in compliance with the current regulations in a given country would not be of interest to a company or industry in that country even if it solved a major problem since it would not allow its implementation.

It often happens that these aspects are not considered in the research design. Although it is ideal to consider regulatory aspects from the beginning, if that does not happen, it is important to analyze them as soon as possible. Not considering them may imply the need to repeat some experiments or even render the technology transfer unfeasible.

In this analysis, it is important to consider that while researching, there may not be relevant national regulations, but there could be international regulations to attend to for those countries. For example, the regulations of the US Food and Drug Administration (responsible for regulating food, drugs, cosmetics, medical devices, biological products, and blood derivatives) are often different and much stricter than the regulations of equivalent entities in Latin America and even different in many cases from the regulations of the equivalent European entity.

Financial Aspects

The financial aspect is crucial, so it is necessary to know how much funding is required, both to reach the next milestone of value creation for development or commercial purposes and to complete its development.

There are various financing options available, so it is important to identify and clearly understand what they allow to be financed. It is also relevant to consider, in the face of different financing options, the obligations or implications that each one entails.

A good strategy should consider continuity in the different financing stages, and compatibility between the various sources in terms of rights granted to third parties, involvement of companies or others.

It is very relevant to consider that the commitments made with previous financing are compatible with new financing; each financing brings with it a series of commitments that must be taken into account when applying for it.

Many times, the first customers will be potential investors. To sell them the idea, it is necessary to convince them that it has a competitive advantage that will remain valid for several years. Investors will contribute if the path to start a pilot is clear and short and will only make a bet if the route and time to complete the pilot are clearly defined, as scaling is where business opportunities arise, along with an exit option for the investor.

References

García, C., Moncada, N., Dussaillant, G., Soto, W., Silva, P., Ogno, N., Bravo, B., Martinoli, S., & Montalva, T. (2014). *The entrepreneurship guide to jump high.* https://www.jumpchile.com/wp-content/uploads/2018/06/LA-GUIA-DEL-EMPRENDIMIENTO-PARA-SALTAR-ALTO_digital.pdf

Gruber, M., & Tal, S. (2017). *Where to play: 3 steps for discovering your most valuable market opportunities.*

Ibañez, J. M. (2014). *Technology readiness levels: TRLS: An introduction* (pp. 165–171).

Osterwalder, A., Pigneur, Y., Bernarda, G., & Smith, A. (2014). *Value proposition design.* Wiley.

Sethi, A. (2016). From science to startup. In *From science to startup: The inside track of technology entrepreneurship.* Springer International Publishing. https://doi.org/10.1007/978-3-319-30424-3

3 Intellectual Property

Abstract

This chapter addresses the fundamental aspects of intellectual property, a set of legal rights designed to protect human intellectual creations, such as inventions, literary, scientific, or artistic works. Two key international agreements are highlighted: the Paris Convention for the Protection of Industrial Property and the Berne Convention for the Protection of Literary and Artistic Works. These agreements establish "common standards" with the aim of ensuring fair treatment and protection of intellectual property in different countries. The protection of intellectual property pursues two main objectives: safeguarding the rights of creators and ensuring society's access to the results of these creations. In addition, it seeks to foster creativity and promote the application of inventions and works in goods and services that solve specific problems. Proper management of intellectual property can have multiple benefits, including promoting knowledge exchange, stimulating economic growth, and generating employment. Not adequately protecting intellectual assets can lead to misappropriation, loss of economic benefits, and use contrary to the original interests of the creations. Therefore, it is crucial to adopt effective measures to ensure the protection of intellectual property and avoid these risks.

Keywords

Intellectual property • Copyright • Patents • Technology transfer • Protection strategy

Key Aspects of Intellectual Property

Intellectual property is a set of legal rights that protect human intellectual creations, such as inventions, literary works, scientific or artistic works, granted by a specific country and governed according to the legislation of each country.

A. Ossa, *From Research to Market*, https://doi.org/10.1007/978-3-031-71340-8_3

It is essential to consider that intellectual property protects the product of the human mind or intellect, and therefore, those who can create or invent are natural persons and not legal entities, such as companies, universities, or other organizations. Furthermore, intellectual property does not protect ideas but the materialization of those ideas. The concept of "copying an idea" is meaningless since ideas are not considered appropriable or copyable, at least formally.

The promotion of intellectual property has a long history and dates back to the late nineteenth century. On March 20, 1883, the Paris Convention for the Protection of Industrial Property was established (WIPO Database of Intellectual Property Legislative Texts, 1979), where the relevance of these matters was first recorded. This convention establishes that intellectual property has a "national treatment," meaning "... the Contracting States shall grant to the nationals of the other Contracting States the same protection as they grant to their own nationals." It also agrees on the "right of priority," meaning that "the first application for a patent of invention or registration of a trademark filed in one of the Contracting States, the applicant may, for a certain period, request protection in any of the other Contracting States." Finally, the convention establishes "common standards" to which all Contracting States must adhere. Although the agreement is over a hundred years old, many of its rules are still in force, and others have had to be modified. Thus, the convention was formally revised in Brussels in 1900, in Washington in 1911, in The Hague in 1925, in London in 1934, in Lisbon in 1958, and in Stockholm in 1967, and finally amended in 1979. Currently, more than 170 countries worldwide have signed the agreement.

Three years after the Paris Convention was established, on September 9, the Berne Convention for the Protection of Literary and Artistic Works was established (*WIPO,* 2020), which, like the first agreement, is based on three basic principles. The first is national treatment, which means that "works originating in one of the Contracting States must be the subject, in each and every one of the other Contracting States, of the same protection as they grant to works of their own nationals." The second is the principle of *automatic* protection, which determines that "protection shall not be subject to the fulfillment of any formality." The third is the principle of *independence* of protection, according to which "protection is independent of the existence of protection in the country of origin of the work." In addition, the agreement established the type of work, the rights conferred, and their duration. After the 1886 agreement, it was revised in Paris in 1896, and in Berlin in 1908, completed in Bern in 1914, and revised again in Rome in 1928, in Brussels in 1948, in Stockholm in 1967, and in Paris in 1971, finally being subject to amendment in 1979. Currently, more than 170 countries have also signed this agreement.

The World Intellectual Property Organization, WIPO, administers both agreements. This organization is part of the United Nations, and was created on July 14, 1967, in Stockholm. It is currently headquartered in Geneva, Switzerland, and has more than 190 member countries; its objective is to promote the use and protection of works of the human intellect.

On the other hand, there is the World Trade Organization, which was created on January 1, 1995, replacing its predecessor, the General Agreement on Tariffs and Trade (GATT), which dates back to 1948. The current institution is composed of more than 160 members, whose objective is to organize for the opening of trade. It has an agreement on Trade-Related Aspects of Intellectual Property Rights (TRIPS) (*WTO—Intellectual Property—Home*, n.d.), which seeks to achieve an appropriate balance between the long-term social objective of providing incentives for future inventions and creations, and the short-term objective of allowing the use of existing inventions and creations. Precisely, Article 7 of the agreement establishes the objectives:

> The protection and enforcement of intellectual property rights should contribute to the promotion of technological innovation and to the transfer and dissemination of technology, to the mutual advantage of producers and users of technological knowledge, and in a manner conducive to social and economic welfare, and to a balance of rights and obligations. (*WTO—Intellectual Property—Home*, n.d.)

Therefore, although intellectual property is governed according to the legislation of each country, there are international agreements that regulate and try to standardize national regulations.

Why Protect?

Countries have two main reasons for legislating on intellectual property. First, they seek to protect the legal rights that people have over their developments and, at the same time, protect society as a whole so that it has access to these results. Second, they aim to stimulate creativity and encourage the application of these new inventions and creations, transforming them into goods and services that solve concrete problems. Clear rules that promote trade exchange are needed in order to contribute to the economic and social development of countries.

As intellectual property constitutes a series of rights that have been codified into law, it allows for commercial exchange; that is, intellectual property rights can be bought, sold, leased, or destroyed, since the government grants an exclusive and excluding right for a certain period. In other words, it grants the owner of said intellectual asset a monopoly right to exploit its development.

The management of intellectual property can bring multiple benefits. It serves to promote the exchange of knowledge through information that must be made available, e.g., open patent databases, which can encourage technological development. Additionally, it serves to stimulate economic growth; if an intellectual property asset is successfully transferred, it could generate new jobs and even new industries.

Another way to highlight the importance of intellectual property is to understand what happens when an intellectual asset (e.g., a research result), is not protected. It could happen that a person who has not been involved in the development appropriates that result, perhaps due to inaction or simply the lack of knowledge on the part of the research team. This could bring about the inability to use the result or cause the use of such results to be controlled by a third party. However, in the patent process, there is the opposition or nullity process, which could reverse this problem. In the process of granting a patent, it is possible for any affected party, within a certain period, to oppose the granting of the patent, alleging the lack of any of the requirements demanded by law; the opposition could imply the nullity of a patent in whole or in part (*Patent Quality, Including Opposition Systems Opposition and Nullity Procedures*, n.d.).

There is also the potential that someone other than the research team and the institution where it was developed obtains economic gains with the research result since, even without having developed it, they will have the possibility to use and commercialize it. If the intellectual asset is not protected, it will be understood that anyone would have the ability to use it to create a product, and someone might be willing to pay for that product.

Another possibility is that a third party uses the unprotected result, which would discourage any public or private funding for research that overlaps with the improperly appropriated results. In that case, it would not be possible to offer any other interested party any improvement or new development linked to the original research result.

One of the most complex situations arising from not protecting an intellectual asset is that a third party could use it for purposes contrary to the initial interest for which it was conceived, preventing the research result from becoming a good or service. Alternatively, by offering it at an excessively high cost, access barriers would be imposed on those who need the product most.

Finally, it could happen that no one uses or exploits the research result. In that case, the circulation of the result would only occur within the scientific community, causing the loss of the value of that knowledge in terms of social impact, as it would never be possible to access the good or service.

Therefore, if adequate protection of the intellectual asset is achieved, in addition to recognizing authorships and benefiting society with the generated knowledge, it could generate a development that solves a specific market problem, can be commercially exploited, and encourages a third party to invest in the research result. In cases where the technology effectively becomes a product and reaches the market, this ensures that competitors will not be able to use or commercially exploit the same technology or a very similar one, generating a competitive advantage for the investor.

It could be said then that intellectual property is wealth. For example, Apple, the most valuable brand worldwide in 2022, according to Best Global Brands, has an estimated value of 355 billion dollars (Interbrand, 2021).

Classification of Intellectual Property

Intellectual property is classified according to the protection mechanism and, in general, is quite similar in all countries of the world, especially in those that are subscribed to international treaties [1]. However, there are variations in the denomination of the categories, as well as in the organization that manages it in each country and even in the years that the protection lasts.

In many countries, intellectual property is divided into two main groups. The first group is called "copyright and related rights." It is usually referred to as copyright and is also known as intellectual property, which makes it confusing for the user, as it bears the same name as the general classification. Copyright is divided into moral and economic rights, and related rights, which consider the rights of the performer, executable rights, and phonogram producer rights.

The second group is called "industrial property" and considers the following: patents, which can be invention patents, utility models, layout designs, or topographies of integrated circuits; design patents, which can be drawings or industrial designs; trademarks, which can be certification, collective, sound, slogan, commercial establishment, industrial establishment, mixed, figurative, and denominative trademarks; internet registrations or domain names; and trade secrets or industrial secrets. In some countries, geographical indications, appellations of origin, and plant varieties are also included in this category.

Copyright and Related Rights

Copyright

Copyright is a constitutional guarantee that ensures all people the freedom to create and disseminate arts and science and safeguards, in the particular case of authors, a set of rights of a patrimonial and moral nature over the works that are the product of their creative activity (Chubretovic, 2013). It also protects the copyright holders, that is, authors, surviving spouse, heirs, assignees, and legatees.

Copyright exists from the very moment of creation, and therefore, it is not necessary to register the work in any registry. However, it is very useful and advisable to register it, as it serves to facilitate proof that one is the holder of the right, as well as establish the date of creation. It is also useful as a commercial strategy, but it is not a requirement to establish it. Therefore, copyright protection does not require any formality and is automatic, so if an author does not register their work, they are equally protected by law and have all the powers derived from copyright (Chubretovic, 2013).

This right protects the original expression of ideas, not the ideas themselves. Protection is valid worldwide but is governed by the laws of each country, with variations, especially in the duration of the right. The Berne Convention guarantees national treatment, that is, "works originating in one of the Contracting States shall be subject, in all and each of the other Contracting States, to the same protection as they grant to the works of their own nations" (WIPO, 2020). In most cases, the right lasts up to fifty or even seventy years after the death of the last of the authors. After this period, the work becomes part of the common cultural heritage, so its subsequent use does not require any authorization.

Works that can be protected under this mechanism include books, brochures, magazines, lectures, speeches, newspapers, photographs, engravings, architectural models, sketches, maps, plans, paintings, drawings, illustrations, adaptations, translations, authorized transformations, videograms and slide shows, software, mobile device applications, databases, musical works, theatrical works, and recordings.

An easy way to exemplify this is to make a comparison between things that are protectable and those that are not. For example, the idea that inspires a choreography is not protected, but a choreography is; a musical style is not protected, but a song is; a legend is not protected, but a book of mythology is; the idea for a movie plot is not protected, but the script is; the idea of a lottery game is not protected, but the computer program created and used for the game is (Chubretovic, 2013).

Moral Rights

Moral rights correspond to "certain measures that the author can take to preserve their personal connection with their work" (WIPO, 2016). This includes the recognition by law of the author's *paternity* over the work they created and the opposition to deformations of the work, among other measures. In this sense, moral rights over a work only correspond to the author of the work and not to the one who acquires it later; that is, it is an inalienable, non-transferable, and personal right that cannot be sold, assigned, or transferred, except by succession due to death (Chubretovic, 2013).

Patrimonial Rights

The economic rights correspond to the right to economically exploit the work and give its holder the possibility to publish, reproduce, adapt, perform, or distribute it, as well as to transfer, wholly or partially, their rights or authorize the use of the work by third parties (Chubretovic, 2013). The author may have both moral and economic rights, but often, the author transfers the economic rights to a third party, so they maintain the moral rights and may even receive income if agreed upon with the new owner of the economic rights.

Related Rights

Related rights are those granted by law to artists, performers, and executors to allow or prohibit the dissemination of their works and receive payment for the public use of their productions (Chubretovic, 2013).

Performer's Right

According to Chubretovic (2013), it is the "right of artists, performers, and executors to allow or prohibit the dissemination of their productions, and receive remuneration for the public use of them, without prejudice to those corresponding to the author of the work."

Executor's Right

The executor's right is the right of those who use a phonogram or sound recording on special supports, which allow their reproduction, such as vinyl, cassettes, compact discs, mp3, or any other format, or reproductions of them for public dissemination, provided that a fee is paid to the artists, performers, or executors and the producers of those phonograms (WIPO, 2016).

Phonographic Producer

It refers to the right to authorize or prohibit the fixation or recording of broadcasts and reproduction by broadcasting or television organizations. In addition, the law protects the related rights of artists, performers, executors, phonographic producers, and broadcasting organizations (Chubretovic, 2013).

Industrial Property

Industrial property is understood as a set of exclusive rights granted by a country to use or exploit patents, designs, trademarks, geographical indications, appellations of origin, and, in some countries, plant varieties in a specific territory, which arise from their registration carried out through formal and substantive examinations. Different treatment is given to internet registrations and trade secrets, which are excluded from the analysis below and will be analyzed later, due to their particularities.

One of the characteristics of industrial property is what is called "territoriality" as industrial property rights only have effect in the country where they were requested and granted. In this sense, it is highly recommended, in most cases, to apply for the right in all countries where the technology will be produced and exploited.

Another characteristic of industrial property is its temporality, as the protection granted generally has a limited duration of years, and then the protected object is considered public domain.

It is also important to understand the difference between the right holder and the inventors. The inventor is the one who makes the intellectual contribution to development and can only be a natural person. The right holder or applicant is the one who has the rights over the invention. This can be the same inventor, the inventor's employer, the one who has contracted the services, or the one to whom the inventor or their employer has assigned those rights, whether a natural or legal person.

Invention Patents

Patents provide exclusive rights granted by the government for protection, which prevent third parties from using them without consent and, therefore, allow the holder to use and exploit an invention. If it is decided not to directly exploit the patent, the rights can be sold or transferred to another person or company in order to commercialize it under certain conditions.

In order for a patent to be granted, the invention must meet three fundamental requirements. First, it must be novel at a global level, meaning that it did not exist before in the state of the art. The state of the art is everything that has been disclosed or made accessible to the public, anywhere in the world, through a tangible publication, sale, or marketing, before the day of filing the application, which reveals an equal development. Second, it must have an inventive level, meaning that the invention cannot be obvious or evidently derived from the state of the art for a person normally skilled in the corresponding technical field. Finally, it must be susceptible to industrial application, meaning that it can be produced or used in any type of industry, whether manufacturing, craftsmanship, mining, agriculture, or others.

The protection granted by the government is territorial, meaning it is only at the national level and for a single period of twenty years from the date of filing the application (WIPO, n.d.-b). For this to occur, the patent applicant must make their invention public through a patent application so that an expert in the field referred to by the application can reproduce the invention. In exchange for publishing their invention, the applicant obtains a right that allows them to prevent others from any form of commercial exploitation of the invention within the territory where the patent is obtained and for the period established by law (Ortiz & Enric, 2010).

Invention patents can protect products, such as compounds, compositions, devices, materials, structures, and apparatuses, among others. It is also possible to protect methods, such as a process, as well as uses or second uses, and systems.

Confidentiality is important in the patent application process, as it is essential to meet the novelty requirement. The disclosure of the technical details of the new development through any means, such as web pages, videos, reports, scientific articles, talks, social networks, or other mechanisms, even by the same team of inventors, may prevent the granting of the patent.

Another key aspect of the patent application process is the search for prior art, also known as the search for the state of the art. This consists of a report that details and analyzes the documents closest to the development to be patented, to determine if they may interfere with the novelty and inventive levels. It is important to perform this analysis before filing a patent application to assess the patentability potential of the new development and to properly focus the drafting of the patent.

Many researchers anticipate the dilemma of whether to publish a scientific paper or initiate a patent application process; however, this dilemma need not exist, as both actions are complementary and not mutually exclusive. The important thing is the order in which they are done so as to fulfill the necessary requirements. It is important to file the patent application before publishing or disclosing it in any way; both things can even be done on the same day (as long as the filing is done prior to publication), as there will be no problems with publishing after filing the application.

Patent applications must be filed with the government offices dedicated to these matters in the country where protection is sought. In addition, there are two international conventions that establish the rules for patenting the same invention in different countries: the Paris Convention, and the Patent Cooperation Treaty (PCT).

Utility Models

As defined by some countries (Utility Model, n.d.), a utility model is an industrial property title that, like patents, protects inventions but with less creative value or non-radical innovation. Utility models generally apply to inventions of lesser technical complexity, which is why they are known as "small patents," "utility innovations," or "short-term patents."

The utility model protects objects that, due to their shape, structure, or the configuration of their elements, perform a new function, meaning that the change in shape solves a new technical problem. This category exists only in some countries. For example, a hammer with an incorporated device for removing nails was protected as a utility model, as it previously only served to hammer and now also to remove nails.

Utility models do not need to meet the inventive step requirement; they only need to meet the requirements of novelty and industrial application. However, a patent will not be granted when the utility model presents only minor or secondary differences that do not provide any discernible utility features compared to previous inventions or utility models (*Utility Model*, n.d.).

In general terms, the following are considered to be utility models: instruments, tools, devices, and objects or parts thereof, with differentiating factors, both in their external appearance and in its operation. It must provide utility; that is, it offers a function, benefit, advantage, or technical effect that it did not have before. The duration of protection for utility models is shorter than that of invention patents. Usually, the protection period is ten years, counted from the filing date of the application.

Integrated Circuit Layout

Integrated circuit layout schemes or topographies are three-dimensional arrangements of the elements that compose them. The integrated circuit is a device in which certain elements with electrical functions, such as transistors, resistors, capacitors, diodes, or others, are mounted on a common substrate. These components are connected in such a way that the integrated circuit can control the electric current, and they are arranged according to the function they are going to perform. The protection is non-renewable and lasts ten years, counted from the date of filing the application for registration or the first commercial exploitation anywhere in the world (*Layout Schemes or Topographies of Integrated Circuits*, n.d.)

Industrial Drawings and Designs

Any three-dimensional form, associated or not with colors, is considered to be an industrial design. It is any industrial or artisanal item that serves as a pattern for the manufacture of other units and is distinguished from its similar items, either by its shape, geometric configuration, ornamentation, or a combination of these; these characteristics must give it a special appearance perceptible by sight, in such a way that it results in a new physiognomy. Examples would include the design of a smartphone or a chair (*Designs*, n.d.).

Unlike industrial design, which is three-dimensional, industrial drawing is two-dimensional. Patterns on fabrics, textiles, or any laminar material are included among the items that can be protected as industrial drawings, provided they meet the condition of novelty. An industrial drawing protects the arrangement of lines and colors of a pattern, not the object to which it is applied. For example, if a fabric is arranged on an armchair, the protection does not include the armchair. Industrial drawings and designs only need to meet the requirement of novelty, not the inventive step or industrial application, as in the case of invention patents. The duration of registered protection varies from country to country, but it is usually ten years (*Designs*, n.d.).

Plant Variety Registration

This protection mechanism considers a set of rights aimed at promoting and encouraging the creation of new plant varieties (understood as a new set of plants of a single botanical taxon of the lowest known rank), which are distinct, homogeneous, and stable. In this category, agricultural seeds and plants, fruit, and ornamental varieties are protected.

The protection period for agricultural and ornamental species varies from country to country, but it generally extends for fifteen years, and for fruit species, for eighteen years. The breeder's right to a new plant variety grants its holder the exclusive right to multiply, produce, sell, market, offer, import, or export it.

Trademarks

A trademark is any sign capable of graphic representation that can distinguish products, services, or commercial or industrial establishments in the market (Schmitz Vaccaro, 2012). The sign must be distinctive and not descriptive, and it must not mislead or deceive the user. When it is said that it cannot be descriptive, it means that, for example, the trademark *agua* cannot be registered to sell bottled mineral water, as it is descriptive of the product it contains and not distinctive as *Agua Veri* (one of the most famous water brands in Spain) could be, according to the Organization of Consumers and Users of that country.

Trademarks can consist of one or more words, such as GOOGLE, or a combination of letters and numbers, known as a wordmark, such as 7 ELEVEN. They can also be composed of figures, images, color combinations, symbols, or drawings without incorporating words, letters, or numbers; these are called figurative marks, as is the case with APPLE's *apple*. When they contain elements of wordmarks and figurative marks, they are referred to as composite marks. Finally, those containing auditory signs are known as sound marks, such as INTEL's *bong* (as it is often called).

Advertising or promotional phrases can also be registered, provided they are attached to a registered trademark of the product, service, or establishment for which they will be used.

Industrial establishment trademarks are those that seek to protect the name of a particular company that produces certain goods. On the other hand, commercial establishment trademarks are those that aim to protect the name of a particular company that trades certain products.

Collective trademarks, on the other hand, are those that allow distinguishing the origin, material, manufacturing method, or other common characteristics of the goods produced by the members of an association; this allows differentiation in the market of products and services from those produced or provided by third parties who are not part of said association or group. This trademark is linked to the denomination of origin or geographical indication, and the holder is the association of producers of the area.

On the other hand, certification marks correspond to those intended to be applied to products in order to accredit certain characteristics by duly authorized persons or entities.

Established in 1957 in France, the Nice Agreement regulates trademarks internationally (*Nice Agreement Concerning the International Classification of Goods and Services for the Purposes of the Registration of Marks*, n.d.). As the name suggests, it aims to regulate everything related to the international classification of goods and services for trademark registration. To facilitate the agreement, the Nice Classification was defined, establishing the international system for classifying goods and services, which is necessary when defining a trademark for registration.

The different products and services are categorized into 45 classes: classes 1–34 include products; and classes 35–45 include services. The idea of this system is to specify and limit the scope of the right, determining the products or services covered by the trademark and, at the same time, unifying the classification system worldwide. The Nice Classification is updated every five years.

Another international agreement governing trademark registration is the Madrid Protocol. Administered by the World Intellectual Property Organization, it has been in force since 1996. It offers trademark holders the possibility of protecting their trademarks in several countries through the simple submission of an application made directly to their own national trademark office (*Madrid Protocol (International Trademark Registration)*, n.d.).

In several countries, the initial registration of a trademark lasts ten years and is renewable for ten-year periods indefinitely, and the protection granted by the trademark is territorial.

Geographical Indication and Denomination of Origin

A geographical indication is a distinctive sign that identifies a product as originating from a territory when its quality, reputation, or other characteristic is fundamentally attributable to its geographical origin; for example, agricultural products that have qualities derived from their place of production, extraction, or cultivation, and are subject to specific local factors, such as climate and terrain (WIPO, n.d.-a). Geographical indications are of indefinite duration as long as the factors that led to their recognition are maintained.

The denomination of origin is also a distinctive sign that identifies a product as originating from a territory when its quality, reputation, or other characteristic is fundamentally attributable to its geographical origin; however, it also takes into consideration other natural and human factors that influence the characterization of the product. Like geographical indications, denominations of origin are of indefinite duration as long as the factors that led to their recognition are maintained (WIPO, n.d.-a).

Within this category are Mexico's tequila, France's champagne, and Colombia's coffee.

Internet Registry

As established by the World Intellectual Property Organization, internet registries or domain names are internet addresses expressed simply for user convenience. They translate the internal protocol addresses or IP addresses of each asset on the network into memorable and easy-to-find terms. Although they were designed to allow users to easily locate information, domain names have acquired greater commercial importance (*First WIPO Process on Internet Domain Names: Archive*, n.d.).

A domain name registration application is valid when the respective fee is paid, lasts for a specified time and is indefinitely renewable. There is an international organization related to internet domains called the Internet Assigned Numbers Authority, IANA. It is responsible for the global coordination of domains and delegates the administration and assignment of domain names to different entities worldwide.

Trade Secret

A trade secret is any knowledge about products or industrial processes that is kept secret and provides its holder with an improvement, advancement, or competitive advantage (*INAPI*, n.d.).

For information to be considered a trade secret, it is essential that it be confidential. However, there is no registration procedure for the company or person to maintain the secret. In addition, it must have commercial value, and safeguard measures must be in place. In this sense, it has a validity for as long as the information that feeds the secret remains confidential.

An example of a trade secret is the formula for a famous soda, which is known by only a few people.

Protection of Computer Programs

The protection mechanism for software or an app has evolved over time and has multiple protection mechanisms. In general, they are protected by copyright and under certain conditions by patents.

In 1966, software protection began through copyright in the United States. By the late 1970s, companies that marketed software began to establish license agreements with users, setting certain conditions and usage restrictions. In 1980 the protection of computer programs through copyright was established by law, and after this many countries began to regulate their protection.

Software could be protected by patents if it produces a technical effect, or when it is part of the hardware, which is known as an embedded system, such as machines controlled by computer programs in manufacturing processes or control procedures. It is important to note that to protect software through a patent, it is not necessary to present or disclose the source code.

Some argue that patenting computer programs would restrict the competitiveness of the sector because producing them often requires the cumulative and sequential development of third-party works and their reuse, which would be incompatible with the need to guarantee the existing interoperability between program components, systems, and existing networks (World Intellectual Property Organization, n.d.).

The most commonly used protection mechanism is copyright and as the law indicates, it is protected from the moment of its creation. In this case, the copyright holders are the people who produce it, but if a person commissions the production of a computer program from a third party, the one who hires will be the copyright holder unless otherwise agreed.

Protection Strategies

The protection of intellectual property is a means and not an end in itself. In this sense, the protection strategy is a necessary plan that allows obtaining value from the protected intellectual assets.

One of the objectives of protection is to give the asset exclusivity in the market and not just give it value as a technical solution. There are many reasons to protect, including to entering new markets, adding value to the company, or preventing copying of the development. These reasons are not mutually exclusive. However, it is essential to be clear about why it wants to protect, as it is a crucial factor in outlining the protection strategy.

To develop an appropriate strategy, it is essential to search and analyze what exists in the world regarding this, for example, by analyzing existing public databases on trademarks, patents, or others. This also ensures that no existing intellectual property assets are being infringed.

If the protection mechanism is through patents, it is crucial to ensure that the information about patentable inventions is not shared with third parties or published before submitting a patent application. The same applies to industrial secrets, for which it is highly advisable to sign confidentiality agreements when negotiating and sharing information with associates in order to protect trade secrets.

It is also necessary, when carrying out joint research and development projects with other companies or research institutes, to ensure that it is clear who is the owner of the potential intellectual property generated by the project.

One of the relevant factors to understand is that intellectual property rights are not mutually exclusive; therefore, an integral protection strategy can and should be designed to generate more robust protection since a single creation can be protected by different rights according to its characteristics.

If, for example, a new food product is developed, such as a cereal bar based on a new variety of corn, its different elements can be protected by different rights: it would be possible to protect the method of manufacturing the corn bar by patents; he product name and label logo could be registered through a trademark; the bar design could be registered by industrial design, and the corn plant by registration of plant varieties; copyright would protect the wrapper design; and the formula for making the product would be protected by industrial secret.

Therefore, it can be said that the management of intellectual property is essential to protect and recognize the creator and the creator's legal rights to decide what to do with their intellectual properties. The mechanisms for protecting intellectual property rights safeguard the use of the different aspects related to development and are powerful tools for generating value in society and the market.

Note

1. In this book, some definitions and classifications that apply to most countries will be provided; however, there may be some variations depending on the legislation of each country ↑.

References

Chubretovic, T. (2013). *Intellectual property and copyright guide cultural legislation.* Chile. http://diseno.uc.cl/wp/wp-content/uploads/2015/08/guia-propiedad-intelectual.pdf

Designs. (n.d.). Retrieved January 5, 2021, from https://www.inapi.cl/patentes/tipos-de-patentes/disenos

First WIPO Process on Internet Domain Names: Archive. (n.d.). Retrieved January 5, 2021, from https://www.wipo.int/amc/en/processes/process1/report/summary.html

INAPI. (n.d.). Retrieved January 5, 2021, from https://inapi.cl/glosario

Interbrand. (2021). *Best global brands—2021 (Interbrand) | Ranking the brands.* Ranking The Brands. https://interbrand.com/best-brands/

International, O. (1979). *Nice agreement concerning the international classification of goods and services for the registration of marks, 1957* (pp. 1–9). https://www.wipo.int/treaties/en/classification/nice/

Layout Designs or Topographies of Integrated Circuits. (n.d.). Retrieved January 5, 2021, from https://www.inapi.cl/patentes/tipos-de-patentes/esquema-de-trazados-o-topografias-de-circuitos-integrados

Madrid Protocol (International Registration of Marks). (n.d.). Retrieved January 5, 2021, from https://euipo.europa.eu/ohimportal/es/madrid-protocol

Ortiz, I., & Enric, E. (2010). *Best practices guide for patent information search.*

Quality of Patents, Including Opposition Systems Opposition and Nullity Procedures. (n.d.). Retrieved December 30, 2020, from http://www.wipo.int/scp/en/revocation_mechanisms/

Schmitz Vaccaro, C. (2012). Distinctiveness and use of trademarks. *Chilean Law Review, 39*(1), 9–31. https://doi.org/10.4067/S0718-34372012000100002

Utility Model. (n.d.). Retrieved January 5, 2021, from https://www.inapi.cl/patentes/tipos-de-patentes/modelo-de-utilidad

WIPO. (2016). Understanding copyright and related rights. *World Intellectual Property Organization, 2*, 1–40. https://www.wipo.int/edocs/pubdocs/en/wipo_pub_909_2016.pdf

WIPO (2020). Database of intellectual property legislative texts. *Paris Convention for the Protection of Industrial Property*. 1–20. https://www.wipo.int/treaties/es/ip/paris/

WIPO. (n.d.-a). *Geographical indicator*, 2023. https://www.wipo.int/geo_indications/en/index.html

WIPO. (n.d.-b). *Patents*, 2023. https://www.wipo.int/patents/en/

WIPO Database of Intellectual Property Legislative Texts. (1979). *Paris convention for the protection of industrial property* (pp. 1–20). https://www.wipo.int/treaties/es/ip/paris/

World Intellectual Property Organization. (n.d.). *Computer programs and business methods*. Retrieved January 5, 2021, from https://www.wipo.int/patent-law/en/developments/software.html

WTO ¦ Intellectual Property—Homepage. (n.d.). Retrieved December 30, 2020, from https://www.wto.org/spanish/tratop_s/trips_s/trips_s.htm

Technology Commercialization

4

Abstract

This chapter analyzes the mechanisms of "technology commercialization," such as licensing and the transfer of intellectual property assets. To succeed in this process, it is crucial to understand the characteristics of the technology, competitors, regulations, business risks, and economic conditions. The ability to design business strategies, assess risks, and negotiate plays a decisive role. Licensing agreements can take the form of transfer or license and cover different intellectual property rights. The decision to commercialize is based on the valuation of the technology and the reduction of associated risks. Finding interested parties, whether patent holders in the same field or companies with complementary solutions, is crucial. Preparation prior to negotiation is essential, considering factors such as technology, customer, market, patents, and research. Knowing the licensee and understanding the technology are fundamental aspects of achieving success in negotiation. A licensing agreement establishes the conditions for the third party to use and exploit the technology. Some key aspects of the content of the license, the type of license, the scope of application, the territory, the duration, the consideration, the termination, improvements, sublicensing rights, milestones, and annual reviews. These aspects are fundamental to establishing the rights and obligations of the parties involved and ensuring a beneficial and equitable relationship.

Keywords

Technology commercialization • Intellectual property • Business strategies • Negotiation • Licensing agreement

A. Ossa, *From Research to Market*, https://doi.org/10.1007/978-3-031-71340-8_4

Key Aspects of Commercialization

One of the central objectives of the technology transfer process is the commercialization of technologies, which is defined as the process of licensing or transferring intellectual property assets and their technological packages in order for the technology to reach the market and to obtain monetary gains when possible, or other types of benefits for both the institution and the research team that developed the technology. According to Dr. Enrique Medellín from the Autonomous University of Mexico (Medellín, n.d.), a fundamental element for transferring technologies is that those in charge of this function fully understand the characteristics and advantages of such development; they must know the competitors and substitutes, the legal and regulatory aspects governing technological development, the business risks of the technology, and the economic conditions of the negotiation.

The success of commercialization also depends on the capabilities of those in charge of the negotiation, including their abilities to design a commercial strategy for the technology, their capacity to assess both technical and commercial risks, their ability to carry out the design of an intellectual property strategy and the valuation of the technology; and, undoubtedly, their negotiation skills.

It is essential to understand that technology commercialization is not limited to the licensing of a patent since, in many cases, the research team, and even the technology itself, can play a significant role. One of the most common mistakes is the belief that the technology is worth only what the patent is worth (without considering the other aspects mentioned), or that if there were no patent the technology would have no value.

Another fundamental aspect is that by licensing a technology, the technical and commercial risk of it reaching the market is shared with the licensee, and the expectation is to reduce the risk as much as possible for the university or the R&D center. In this sense, it is important to remember that a licensing agreement must be favorable for both the university and the researchers, as well as for the licensee company since it should not be forgotten that the company is assuming a significant part of the risk.

To carry out this transaction, it is necessary to sign a contract, either a license or a transfer of intellectual property rights. These two types of contracts have different legal implications: a *license* can be compared to the rental contract of a house (that is, the house remains the property of the owner, but it grants usage rights to a third party); a *transfer*, however, is like the sale of a house, where all rights to the asset are lost. The World Intellectual Property Organization (WIPO, 2005) provides a clear reason for granting a license:

> [...] innovations and other expressions of human creativity have become private property and are protected by law through the intellectual property system. As property, they are marketable assets. Licensing, that is, the right that the owner of the assets grants to a third party to use them, while the owner retains ownership of them is an important way to add value to those assets. Licensing generates a source of income, spreads technology to a wider group of users and potential developers, and acts as a catalyst for other developments and commercial exchanges.

Types of Agreement

Assignment Contract

The assignment of intellectual property assets involves a transfer of ownership rights over the asset to a third party in exchange for receiving compensation or consideration, both economic and of other types, as agreed by the parties. Normally, an assignment has a higher value than a license, just as the sale of a house has a higher value than a monthly rental.

A relevant aspect to consider in this type of agreement is that all control over the technology is lost unless some rights are reserved, which are commonly the rights to continue researching and to carry out academic activities with the research results. In the case of research, it is very common to reserve disclosure rights, which allow for publications or presentations at conferences, as long as they do not affect key aspects of the business considered confidential, which must be clearly established in the agreement. It is also important to resolve whether the research rights are commercial or non-commercial. If rights for non-commercial research purposes are reserved. It is implied that any development carried out, whether in the same line of application of the assigned technology or in a very different one, cannot be transferred under any mechanism. This does not occur with rights for research that do not imply restrictions.

The above shows that it is very relevant to analyze when it is suitable to agree on the assignment of a technology. Of course, if the compensation is extremely attractive, this may be reason enough to make the decision. Another criterion that helps determine if it is more convenient or less risky to assign intellectual property is when it is limited and solves a specific problem or points to a particular solution for a company. However, it is very important to consider that such an assignment does not block the research line or future developments of both the institution and the research team.

License Agreement

The license agreement is a legal document by which a natural or legal person grants a third party, who can also be a natural or legal person, the right to use, exploit, develop, and commercialize intellectual property rights. The license agreement is the most frequent instrument for transferring technology, as it implies maintaining ownership of the developments.

Making a good license agreement is not an easy task and depends on multiple intertwined factors such as the characteristics of the technology, the relationship between the negotiators, and the type of business to be carried out, among many others.

In some cases, the license has commercial purposes. This depends on what the institution and the research team pursue. Different intellectual property rights can also be licensed, such as patents, copyrights, software, trade secrets, or even know-how (i.e., *knowing how to do* or *knowing how*), which in some legislatures is considered similar to a trade secret.

A license agreement involves maintaining intellectual property rights, establishing certain conditions (such as ensuring that technology effectively reaches the market), guaranteeing the continuity of the research line, obtaining consideration, or a combination of these and other conditions.

License Option

On some occasions, "license options" are signed instead of direct "license agreements." A license option is a promise to license the research results to a third party, which is generally used when third parties are involved in the development of the technology from the beginning, but the expected results have not yet been achieved. In the case of licenses, the rights to use, exploit, develop, or commercialize are granted from the moment the license is signed.

If a license option is chosen, it is essential to establish the conditions for exercising it, as it can be automatic but with early termination if certain agreed-upon conditions are not met. It is also common to set deadlines for the third party to exercise the option. If they do not exercise their option, the institution is released and can grant the right to another party; however, if no deadline is set and the third party does not exercise their option, the legal transfer of the technology to another party could be blocked.

Contract Negotiation

When to Commercialize?

It is not easy to determine the most appropriate time to license a technology, nor is it certain that such a moment exists. Some argue that it is better to license it as soon as possible, as the technical and commercial risks are shared with others. Moreover, third parties help reduce the risk that the technology never reaches the market. On the other hand, there are those who argue that the moment of the technology's highest value should be sought for licensing, which implies greater benefits for the research developers, although it is possible that this moment never arrives or, alternatively, that the sought-after licensee is not found at the moment of highest value (Sullivan, 1995).

It is well known that technology transfer is immersed in the risk industry; in this sense, the premise that higher risk implies higher value, and vice versa, holds true. One type of risk that can be better managed by developers is technical risk, so theoretically, the more development there is, the lower the risk. In this regard, it is advisable to establish a roadmap that considers key factors, such as progress in research development, intellectual property strategy, regulatory factor analysis, and commercialization strategy. To do this, it is essential to raise resources to continue development, reduce technological risk, and increase the chances of commercialization. This process often begins as soon as intellectual property protection efforts are initiated (or even earlier), and then potential licensees are identified, and negotiations begin.

Searching for Interested Parties

One of the key factors for transferring a technology is finding people or companies interested in it, and there are various characteristics or conditions that can help identify them.

One alternative is to find individuals or legal entities that hold a patent in the same field of application as the technology to be transferred. It follows that if they already have technologies in the same field, they might be interested in learning about new, similar developments.

Another option is to look for companies in the sector that already have a solution in the market or could be developing alternative solutions to the same problem that the technology to be transferred solves. It is natural to think that they might be interested in learning about alternative solutions.

They can also be companies that have product lines possibly complementary to the research result, and that might want to expand their operations.

On some occasions, there are companies that simply seek to support the development of the new technology or even entrepreneurship associated with the technology, as it would be strategic for the industry. An example could be companies in markets that are destined to disappear or whose waste could be inputs for the production of this new technology, thereby reducing their negative externalities.

Preparation

Upon initiating the negotiation of a license, multiple factors must be considered, such as the type of technology involved, the perceived risk of the technology by the client, the current state of development of the invention, the projected cost to bring the transformed technology into a marketable product, the size of the potential market, the profit margin, the robustness of the patent claims or the strength of other intellectual property protection avenues, the likelihood of the patent application(s) being granted, and the estimated cost of the research that led to the invention, among many others (Van Norman & Eisenkot, 2017). This implies that before starting a negotiation, it is necessary to be well-prepared to face it and manage the information of all these factors, which is crucial when establishing the price or even closing the deal.

To conduct an adequate negotiation, it is essential to know and understand the potential licensee (i.e., the type of company and its history); negotiation could be different for a family business than for a publicly traded multinational corporation. Although in both cases, the goal is to maximize shareholder benefits, their motivations or the risk they are willing to take have their own implications.

On the other hand, it is also crucial to know the technology in depth, the initial motivations of the development team, the associated benefits, how it compares to its competitors or substitutes, etc.

In the negotiation of a license agreement, both parties should ask questions not only about their own position but also about the likely position of the other party. Each will be better able to negotiate if they try to understand the other's position. These are key aspects to understand: how relevant the technology will be to the business; the negotiating power of each party; how much time there is to sign the license agreement; who will be on the respective negotiating teams; what are the positions on key issues or aspects of the license; or, what is the negotiation strategy (WIPO, 2004).

Another fundamental aspect when negotiating a license is managing expectations. For example, when a third party negotiates a technology transfer office or a technology broker, it is important not only to keep the inventors well-informed but also to make them understand from the beginning what can and cannot be achieved with the negotiation.

Conditions of a License Agreement

The most important thing in a license agreement is to establish the conditions for a third party to use or exploit the technology. Some of the most relevant aspects that should be considered in any license are the following:

1. License content

In the license content, it is necessary to clearly indicate what is being licensed. In this sense, it is advisable to individualize it; for example, one could refer to a "technology package" that includes the "know-how" associated with the research result and the intellectual property rights that were managed to protect that result, such as patents, copyrights, trademarks, a prototype, etc.

2. Type of license

Another fundamental aspect to be considered in a license agreement is the type of license that can be granted, which can be classified into two categories: exclusive licenses and non-exclusive licenses. When we talk about exclusive licenses, we refer to the fact that all rights are granted to a single individual or legal entity, and therefore, the same rights or part of them cannot be given to another. There are also non-exclusive licenses, which imply that more than one individual or legal entity could have rights of use or exploitation at the same time.

It is important to understand that there are combinations of alternatives, such as exclusivity subject to the scope of application or exclusivity subject to a defined time horizon. Exclusivity can also be granted subject to milestones or alternatives for co-exclusivity, such as territories, scopes of application, and others.

3. Scope of application of the technology

Another relevant aspect refers to the scope of application of the technology. Some technologies may have a limited field of application; for example, a vaccine against a respiratory virus. In some cases, the same technology may be applicable, in different fields; for example, a data analysis technology can be used for decision-making in both the medical and transportation areas.

The scope of a license can also intersect with the type of license (exclusive or non-exclusive), and in this way, determine exclusivity in certain scopes and non-exclusivity in others. In cases where technologies have more than one scope of application, a strategy could be to grant exclusive licenses by scope. For example, it would be possible to grant exclusive rights to a company for medical applications and, at the same time, grant another company exclusive rights for transportation applications.

4. Territory

It is also important to agree in the license document which territories are included within the granted rights. Each country grants intellectual property rights independently. However, the rights to use the technology could be granted in a specific region within that country. If protection is held in more than one country, this allows granting rights with certain characteristics in one territory and completely different ones in another territory in order to diversify business risk. On the other hand, if a company operates in a particular territory, it may not be interested in receiving rights in other territories.

Furthermore, many companies seek rights in all territories. Not necessarily because they operate globally, but to prevent competitors from accessing the same technology in the market.

Some argue that rights can only be granted where proper protection exists, meaning exclusive rights cannot be offered in territories lacking patent protection. However, others argue that it is possible to grant licenses with a worldwide character or with validity for specific territories, even if protection is not held in all territories. The important thing to understand is that in territories without protection, the technology becomes more vulnerable, as it can be copied.

The territoriality of a license can also intersect with exclusivity/non-exclusivity, in addition to the scope of application. That is, an exclusive license could be granted for a particular scope of application in a specific territory, while a non-exclusive or exclusive license could be granted for the same scope or for other scopes in different territories.

5. Duration of the license

The license duration refers to how long the licensee is granted the rights to use or exploit the technology. After that time, the licensee loses the rights.

Some argue that licenses can be granted for a maximum term equivalent to the duration of intellectual property rights; that is, if a patent is licensed, it cannot be licensed for more time than the validity period of the invention patent, which is twenty years. Others argue that they can be granted for a longer time than the duration of the patent, since in some cases a licensed technological package includes *know-how*; however, the technology becomes more vulnerable to being copied. It is also possible to license a technology for shorter terms, which in some cases establishes the possibility of renewal if certain requirements are met, or even automatic renewals. If licensed indefinitely, it is advisable to establish termination conditions for non-compliance with certain previously established clauses, which creates incentives to continue developing and exploiting the technology. However, this also produces a degree of uncertainty for the licensee.

The duration of the license may also intersect with exclusivity/non-exclusivity, and determine a period of exclusive rights and, subsequently, non-exclusive rights. The same applies to scope and territoriality, that is, a license can be granted for a specific period in a specific scope for a particular territory, exclusively or non-exclusively, as well as granting a perpetual license worldwide and for all areas of application, exclusively.

6. License consideration

Another key aspect is the consideration of the license. In exchange for granting the license, the holder establishes the consideration conditions, which can be of different types.

Free: In some cases, no economic consideration is established. However, the fact that no royalty or other payments are required does not imply that certain conditions are not established for this to occur. These conditions could include but are not limited to allowing free access to the end user or in a specific territory; donating products; not paying for the license but committing to market the product according to a specific work plan; committing to provide information to the licensor that benefits the research team in their research; advertising that the licensor has generated the development; or, covering all patent expenses for the duration of the license.

Onerous: In most cases, some type of economic consideration will be required. In this regard and although the parties are free to agree on the type and amount of consideration to be paid, there are some types of consideration commonly used. These may appear in licenses either uniquely or as a combination of more than one of them:

> *Up-front payment*: This is a sum of money to be paid at the time of signing the license. This is generally applied when the technology is in an advanced stage of development, or when there is high interest from the licensee due to the impact of its development.
>
> *Royalty*: This is the most common type of consideration. It consists of a fixed or variable percentage, preferably established on net sales, although it could also be on profits associated with the licensed research result and the related intellectual property. It is generally less than 10% and will depend on multiple factors, such as the state of development, the scope of application, the operational margins of the business, and, of course, the negotiation that takes place. In general, software licenses are those that receive a higher royalty, and biomedical developments correspond to lower royalties. It will also depend on whether the technology is a part and piece of a larger development or is the final product to be marketed, that is, whether it is a monoclonal antibody of several antibodies as part of a vaccine or, rather, corresponds to the development of a vaccine itself. If it is the latter situation, the royalty would be higher than in the first since, in most cases, it is agreed upon the selling price of the final product. This is because it is very complex to audit a company's profits or review its accounting in detail, so easy-to-review processes are established, such as the final invoicing.
>
> Royalties can also be tiered; that is, certain royalties are charged up to a certain number of sales; above those accumulated amounts, a higher royalty percentage is charged. It is common to find between two and four royalty tiers.
>
> *Fixed sums*: These are usually used in cases where there will be no sales on which to charge royalties. Examples include: a process technology or a research tool; when calculating the royalty is very difficult (perhaps because the licensed technology will be incorporated as part of another technology); or, when eliminating uncertainty is preferred.
>
> *Milestone payments*: In some cases, payments, their suspension, increase, or progressive reduction, are subject to the achievement of milestones. Among other things, this could include the start or obtaining of regulatory approvals, granting or rejection of patents, scaling of technology, or entry into other markets.

Generally, in a single license, some of the alternatives seen above are usually combined.

7. Termination

While the natural termination of a license agreement occurs upon the fulfillment of the term established in the agreement document, it may also happen that one of the parties has breached one or more obligations provided for in the license. The following are some of these obligation breaches:

Milestones: Development milestones, scaling, preparatory commercial actions such as regulatory approvals, entry into certain markets, or others may condition the continuity of the license, or its conversion to non-exclusive.

Payments: Non-compliance with the planned payment scheme may also imply the termination of the license.

Other obligations: For example, sending reports on time, lack of payments for expenses associated with processing, maintenance, or defense of intellectual property rights may also imply termination if it has been provided for in the license.

8. Improvements or new developments

It is important to determine whether the licensee can or cannot generate improvements or new developments from the licensed technology, and whether there are royalties associated with their exploitation, as well as who will own these improvements and new developments.

As with all aspects discussed before, rights must be expressly granted; so, if someone who wants to acquire a technology intends to conduct research based on it with the aim of generating new developments or improvements, they must have authorization from the technology owner and original intellectual property rights.

9. Sublicensing rights

Another factor to consider is whether or not sublicensing rights will be granted. This determines whether the company receiving the technology (the licensee) can or cannot pass it on to a third party (the sublicensee) for use or exploitation. It is common for companies to have sublicensing rights, paying attention to some aspects. For example, whether the company is related to the company holding the original rights, it is important to avoid sublicensing solely for the purpose of paying lower royalties. Another important factor is that a sublicense should pay the technology developer between four and five times what is paid for a license; that is, if the royalty for a license is 5%, the sublicense should be between 20 and 25% of sales. This is because when the licensee exploits the product, they must cover production or marketing costs; however, when a sublicensee exploits it, the licensee will receive the full income. For example, if a licensee sells it for 100, and the royalties are 5%, the royalty to the technology developer will be 5.

If the licensee were to sublicense it at 10%, and the sublicensee sells it for 100, the licensee will receive 10, of which 25% (2.5) goes to the technology developer. This makes sense when the sublicensee can mass-produce the technology and, therefore, the income is greater for everyone.

10. Milestones

One of the fundamental objectives of licensing a technology is for it to reach the market and ultimately be in use. In this sense, it is essential to establish conditions for this to happen and, if it is not happening, to have the possibility of granting rights to another third party to make efforts to bring the technology to the market.

One objective mechanism that allows measuring whether the efforts and results are correct is the establishment of milestones, whether technical or commercial. This will allow for early termination if the conditions are not met, and receiving payment for milestones achieved.

11. Periodic reviews

Another key factor is periodic reviews. In general, semi-annual or annual report submissions are established, and a clause is included authorizing audits, mainly when there are doubts about whether the information provided is truthful. It is also determined that if there is an error greater than 5% in the information provided, the party who committed the error pays for the audit. If, on the other hand, the difference is less than 5%, the party who requested the audit pays for it.

12. Reimbursement and payment of patents

A very important aspect to consider is the cost of maintaining patents. In some cases, the licensee pays the institution the direct costs incurred in protecting intellectual property prior to being licensed and, in most cases, assumes the future costs of protection. One of the incentives for the licensee to do this is that in order to keep patents active, it is essential to make the corresponding payments, and they prefer not to run the risk of the licensor ceasing to do so. On the other hand, patents are financial assets, so they have special treatment in accounting. All of the above occurs when an exclusive license is granted. If not, it is difficult to get the licensee to make the payments.

The relevant aspects of a license, such as content, level of exclusivity, duration, territory, and costs, among others, are decisive when negotiating and can be combined depending on the type of license required. For example, exclusivity can be granted in a specific application of the technology. In the case of a genetic improvement technology applicable to multiple crops, rights could be granted for a specific crop (e.g., corn), and the right to other crops (e.g., such as soy or rice) would be reserved for the technology provider's own use, or to license it to another. That is, exclusivity may be subject to the scope of application. Or, exclusivity can be granted subject to a certain period considered appropriate, e.g., for the licensee to recover their investment. Exclusive rights can also be subject to a certain relevant territory for the business and may not be exclusive for other territories. Exclusivity can also be maintained as long as certain pre-established milestones are met, and if not met, it will become non-exclusive. Finally, exclusivity in more territories or areas can also be subject to higher economic considerations.

For a moment, think of each of these conditions as sound that must be regulated through an audio mixing console. Once the sound signals enter, they are combined in various ways to result in a desired audio mix output. If one of the sound signals is too high or too low, the final sound will not be the best, unless an adjustment is made. The same happens with the conditions of a license. If one of them is too *high* or too *low*, it is possible that the contract will not *sound* good. However, with a license if it is necessary to *raise* one of the conditions, it is very likely that it will be necessary to *lower* another one so that the contract *sounds* good for both parties.

In summary, different combinations of these degrees of exclusivity, scope, content, territories, duration, and consideration, among others, can be made in order to safeguard the appropriate use of technology and obtain the expected benefits from its transfer. To define the most appropriate strategy, it is key to understand what each of the actors involved wants to achieve with the transfer of a particular technology and, from there, choose the best strategic combination.

Negotiation

As indicated in the guide "How to Negotiate Technology Licenses" by the World Intellectual Property Organization (WIPO, 2004), the negotiation of a license is a very complex process because it has many facets, and changing one of the variables can affect the others already agreed upon. It will, therefore, be necessary to revisit them each time there are changes. Many of these variables are essential, so it is difficult to ignore one in favor of another. However, it is also true that some essential variables are more important than others, and this will provide room to maneuver when negotiating.

The guide also suggests that it is important to understand that if both parties have essential conditions that are non-negotiable (making it impossible to reach an agreement), it is better to end the negotiation rather than force it. In this sense, it is better not to sign an agreement with completely unfavorable conditions for one of the two parties than to sign it and have it fail in the short term.

One alternative in negotiation is to review the conditions and see if it is possible to change positions. This occurs when new contextual information is included or when access to information that was not available at the beginning is granted, making it necessary to rethink the essential conditions. It is very important not to lose sight of the essential conditions and the business objectives in order to reach an agreement in the negotiation. To do this, it is necessary to prepare very well, understand the essential conditions, and know how much room there is to move between them (WIPO, 2004).

The closing of a successful negotiation will depend on the negotiator's ability to convincingly and clearly explain the objectives of the negotiation and the benefits for both parties at the right time and in the correct way. A negotiation is not a battle; there is no need to raise one's voice or fight. this can damage what is expected to be a long-term relationship. It is likely to be a long time before the licensed technology reaches the market, or the parties are expected to work collaboratively on new versions of the technology, even developing new technologies together. If a license is viewed in isolation from the context and motivations behind the negotiation, it could lead to the conclusion that the negotiation is being poorly conducted.

Therefore, the success of a negotiation will depend on prior preparation and the ability of the negotiators to reach an agreement.

It is advisable to start a negotiation in person, in a relaxed and friendly meeting where both parties show their interests and ambitions regarding the agreement, and expectations can be established, such as the time they expect the negotiation to last. In a first meeting, not all information will be disclosed; it is very common for both parties to withhold some commercial and technical information.

It is advisable to leave the signing of confidentiality agreements for later and only when strictly necessary. Requesting in advance to sign an agreement to keep the information exchanged secret could generate unnecessary friction or ruin the negotiation that is just beginning.

It is important to note that the place where meetings are held influences the negotiation. The host will feel more secure and have greater negotiating power than the visitor. In this sense, it is good to take turns hosting or finding a neutral place if necessary.

In a subsequent meeting, the essential conditions should be discussed, without leaving anything aside but without necessarily seeking to reach an agreement at that time. This clears up matters related to those conditions that are agreed upon, and leaves those that generate more distance for a later meeting.

In order to reach an agreement, in some cases the argument of "usual practices" is used, which is not always convincing since each negotiation is case-specific. In this sense, it is much more advisable to refer to the initial conversations, remember what each party expected from this agreement, and how one or another condition will allow achieving that objective.

It is not advisable to bring lawyers to the first negotiation, but it is important to include them in the negotiation, either by reviewing the agreements or participating in the meetings sooner rather than later.

Once agreements are reached, it is very important that not much time passes before signing the agreement document, even if it is not the definitive license contract.

Signing and Follow-Up

Signing a license contract is just the beginning of a long relationship. Because a business relationship is being initiated, it is recommended that once the license contract has been signed, a symbolic act or celebration with the counterpart be held. The agreement is the beginning and becomes an important guide for what should happen in a complex business relationship based on technology.

The terms of a technology transfer can be simple, but technology licenses are generally more complex and often impose significant conditions, the non-compliance of which can create legal liability and business mistrust. Therefore, they require review by lawyers, which makes the process longer and more complex (WIPO, 2004). Since license agreements have rights and obligations for both parties, it is very important that everyone in both organizations is thoroughly familiar with the license and its terms.

Agreements often have important dates that must be remembered. For example, if one party has agreed to pay the other based on the achievement of certain milestones, these must be verified and collected by the other party. It is also crucial for the company to keep track of deadlines for the delivery of technology prototypes and documentation and to check whether deadlines for research and development are being met. Technology licenses generally involve the payment of royalties. In this way, if you are the licensor, you will need a system to track payments and monitor royalty collection; on the other hand, if you are the licensee, you must keep a record of royalties to be paid and maintain proper documentation. It is advisable to review the agreement, identify such terms, and assign responsibility for tracking each one (WIPO, 2004).

Finally, the license agreement generally has an end date, expiration, or renewal. At that time, it could consult the agreement to see which key terms have been advantageous and which should be reviewed if the license is to be renewed (WIPO, 2004).

References

Medellin, E. (n.d.). *The commercialization of technology* (pp. 1–3).

WIPO. (2004). *How to negotiate a license agreement.*

WIPO. (2005). *Exchanging value: Negotiating technology license agreements* (p. 182).

Sullivan, N. F. (1995). *Technology transfer (Making the most of your intellectual property).*

Van Norman, G. A., & Eisenkot, R. (2017). Technology transfer: From the research bench to commercialization: Part 1: Intellectual property rights—Basics of patents and copyrights. *JACC: Basic to Translational Science*, *2*(1), 85–97. https://doi.org/10.1016/j.jacbts.2017.01.003

5 Science-Based Entrepreneurship

Abstract

This chapter analyzes how science-based entrepreneurship, known as spin-offs, is crucial for the development and transfer of technology to the market. Although they face unique challenges, such as technical and commercial risks, these ventures can offer concrete solutions and contribute to the market. A balanced and multidisciplinary entrepreneurial team, along with the assessment of commercial viability, are fundamental to success. In addition, it is essential to have entrepreneurial capabilities and a team with business experience and commercial knowledge. Factors such as proximity to research centers, access to financing, quality of life, adequate infrastructure, and a favorable business environment also influence these ventures. On the other hand, the importance of the shareholders' agreement and the *exit* for science-based ventures is discussed, regulating the interaction between partners and allowing the monetization of investments.

Keywords

Science-based entrepreneurship • Spin-off • Technology transfer • Entrepreneur • Entrepreneurship • Exit • Shareholder agreement

Key Aspects of Science-Based Entrepreneurship

A technology can be transferred to a company already operating in the market or created to develop, use, or exploit it. The latter is called science-based entrepreneurship or *spin-off*.

Science-based ventures play a key role in both the technical and commercial development of technology. That is why their creation is of great importance in technology transfer processes.

A. Ossa, *From Research to Market*, https://doi.org/10.1007/978-3-031-71340-8_5

Therefore, science-based entrepreneurship can be understood as the formation of a company to use, develop, or exploit a research result created within a university or some institution that develops science, a company in which at least one of the people from the institution that generated the research result participates, or the same institution under the umbrella of which the technology was created.

These ventures can also be called *spin-offs*, which refer to a development that is generated within a scientific organization that is *pushed outward* to give it its own independent life.

Although creating a spin-off is not easy, it is an initiative that has significant advantages for researchers; one is that they can continue their academic activities within the university and, in parallel, carry out the tasks of their venture, which must be previously agreed upon with the institution.

Another motivation for creating a spin-off is that it can involve monetary benefits for both research teams and the institution. This, in turn, leads to the creation of high-quality jobs and stimulate the country's economy.

However, one of the main motivations for embarking on the path of science-based entrepreneurship is to ensure that the developed product or service can effectively reach the market and solve a problem. It is very likely that just the prospect of seeing the initiative materialized is incentive enough to inspire the researcher to take on the challenge.

Before making any decision, researchers must understand that the primary objective of a company is to be profitable, so the decisions they make should pursue financial stability, understanding that research within the company is a means and not an end in itself.

In the publication *Dynamics of Science-based entrepreneurship* (Colombo et al., 2010), it is argued that most science-based ventures do not achieve high growth. However, the technology they develop can contribute significantly to providing concrete solutions to their customers, or it can be transferred and successfully exploited by other companies. That is, although most science-based ventures do not fall into the high-growth category, they can play a very important role in the markets in which they are introduced. There is a significant difference between creating a traditional non-research-based venture and a science-based venture. While both face similar challenges (e.g., market risks associated with understanding their customers well and technical risks related to technological development), science-based ventures may represent the higher risk because these are at the frontier of knowledge. Lubik and Garnsey (2016) argue that they also involve other types of risks, such as those associated with the lack of commercial experience in the entrepreneurial team, the need for high financial resources due to the costly nature of their realization, and divergent objectives on how to advance in the development of science versus the creation of commercial value.

Science-based ventures are not always the best alternative for transferring technology to the market, as they present many challenges and involve high costs, so it is essential to analyze when it is opportune to carry them out. There are some conditions that make the decision more reasonable, such as having a highly capable entrepreneurial team, having created a technological development platform, or having the required financial capital investment.

For the formation of a science-based company, the characteristics of the entrepreneurial team are as important or more important than the technology itself that is intended to be brought to the market, so the interdisciplinarity of its members is crucial to driving the venture. It is essential to have technically knowledgeable people, which is very common in developer teams, but it is also crucial to involve people who have a business or management profile to achieve a balance that allows the venture to move forward. Trust among team members in the real complementarity of their skills and the decentralization of decisions by areas of competence are also key factors for the success of a venture.

On the other hand, the creation of a spin-off makes more sense when there is an innovative technology with high growth potential and a business idea that allows its commercial exploitation. In this regard, one of the aspects to consider when deciding whether or not to set up a science-based venture is whether the technology favors an entrepreneurial model. That is, there are technological development platforms from which multiple other technological developments can be created; these can materialize as various products that meet more than one need in the market or even target different customers.

Let's think, for example, of a technology that allows modeling and designing construction materials, and generating products with different characteristics that can be used for civil works construction. If, in addition, the same technological platform allowed for the creation of materials for medical supplies, it would mean that it is targeting different markets. Because risk would be diversified, the chances of success could increase, and so it would be appropriate to create a spin-off.

Generally, technology development requires high financial capital; public funding is not enough. To achieve private resources, the creation of a venture is needed as a vehicle for financing technological development.

However, it should not be forgotten that an excellent team without technology is not enough to raise capital, and vice versa. Generating a technology-based venture is not always the most appropriate alternative and, therefore, must be analyzed case by case, taking into account multiple variables. Nevertheless, the potential benefits of this path can be enormous.

Considerations of a Science-Based Entrepreneurship

Entrepreneur Profile

In the publication *Dynamics of Science-based Entrepreneurship* (Colombo et al., 2010), the characteristics of the entrepreneur in relation to the success of the venture were analyzed. The study shows, for example, that there is a positive relationship between the success of the venture and the years of university education in economics and management, compared to the years of university education in sciences and technical fields; the latter does not have a significant influence on success when starting a business. It was also observed that previous work experience in the same industry in which the venture is developed has a positive impact on technical work experience. However, the study also points out that the commercial work experience of entrepreneurs does not contribute to growth unless it is associated with the specific technical experience of the industry.

On the other hand, the study *Researching Scientific Entrepreneurship in New Zealand* (Menzies, 2012) compares the characteristics of entrepreneurs and scientists when developing a business. Among the attributes studied are the willingness to take risks, openness to experimentation, and ability to recognize opportunities. Additionally, the study reveals the importance of emotional intelligence and social and interpersonal skills for building and cultivating networks and developing social capital, which is also analyzed to facilitate the success of entrepreneurship.

The willingness to take risks is a topic that frequently appears when studying the attributes required in a scientific entrepreneur. However, there are studies that indicate that it is not just a function of personality but also seems to reflect the social and organizational contexts (Menzies, 2012).

The study also points out that the characteristics required to do science differ from those required to undertake entrepreneurship and are often considered as antagonistic attributes. Even so, many of the traits required by scientists are not inherently different from those required by people working in many other fields, e.g., imagination, self-criticism, diligence, and curiosity.

The study continues by suggesting that, like entrepreneurs, scientists detect opportunities and take risks, although they have less focus or interest in financial aspects. However, one aspect that could be considered different is that scientists are motivated to generate, disseminate, and exchange information, which can be contrasted with the need to maintain confidentiality in business aspects. Hence the challenge is to find the right balance between both. In this sense, scientists do not need to replace altruism with economic benefit but rather consider profit-making to serve their university and themselves, do science, and contribute to the common good.

The study *Effectuation and causation in science-based new venture creation* (Villani et al., 2018) suggests that scientific entrepreneurs seem to rely more on planning than traditional entrepreneurs, who are more inclined towards flexibility. It is argued that because science-based ventures require a long time from the beginning of basic research until the creation of the new company is considered, planning is needed more than flexibility.

The study continues by analyzing the risk aversion of scientific entrepreneurs and concludes that it is related to the information asymmetry about market dynamics and business-related variables.

There are several studies that have demonstrated certain common personality traits among all successful entrepreneurs, including persistence, innovative perspective, high energy level, and propensity for risk and efficiency. In addition, the factors that have the strongest link with entrepreneurial success are high efficacy, the ability to detect and recognize opportunities, high personal perseverance, high human and social capital, and high social skills (Menzies, 2012).

It is also true that many of these capacities or skills can be developed and, above all, complemented with the knowledge of other people, so each inventor has the possibility of becoming an entrepreneur if their motivation is driven by the desire to commercially exploit their invention.

The Team

The entrepreneurial team is essential for carrying out a science-based venture, and a good composition of this team increases the chances of business success. It is important to understand that founding teams cannot be considered as a static concept (Colombo et al., 2010). Throughout the life of the venture, the team composition will likely change, usually complementing it with management and business capabilities, and some team members may stop participating. It is very common for the founding teams of science-based ventures to be unbalanced in terms of being highly experienced in research and development while lacking in commercial functions and applied technological experience.

It is most common for teams to start with a very good technical group, generally available to dedicate the necessary time to the project. It is normal that the technical team does not feel completely confident when starting a business, as the elements required to create a company based on technology are very different from creating the technology itself (Sethi, 2016).

There are two clear and necessary profiles for forming an entrepreneurial team. On one hand, there are the business people who understand that a very solid technology with a competitive advantage is needed to build a company. On the other hand, there are the more technical people who often have a biased view towards commercialization and, in most cases, do not even understand the challenges it entails (Sethi, 2016).

Another difference between scientists and those more oriented towards business is that the former constantly seek to perfect developments to achieve better results. In contrast, those with a more commercial vision focus on bringing the solution to market as soon as possible. Because they are in contact with users, they can identify perceived value for the end customer, market risks, and other challenges for further technology development.

Researchers do it for the sake of research and creating knowledge, which allows for new discoveries. This often makes it difficult to align their intellectual interests with the need to bring their innovation to the market. However, once a technology with a clear commercial advantage is identified, the focus should be on reaching the market as soon as possible. Since technology is nothing without the technical team, this implies that the research team must be motivated to make this their focus as well. Otherwise, a better or cheaper technology, or even an inferior but functional technology, may arrive first (Sethi, 2016).

Commercializing an idea requires a high degree of motivation since each step is fraught with challenges and setbacks, making it critical to align the vision among team members. A conflict could arise if one wants a small business entirely owned by the initial partners while another member wants to pursue global expansion, which will require many external stakeholders, including short-term venture investors and long-term strategic partners (Sethi, 2016).

Therefore, the formation of interdisciplinary and complementary teams are recommended, as not everyone has the same competencies. It is also important to agree on team roles in advance (e.g., who will be in charge of technical leadership and who will be in charge of commercial leadership) and assign decision-making areas based on each member's competencies. Another factor that generally causes problems among teams is the time commitment, so it is essential to establish this clearly from the beginning. Finally, it is crucial to measure team members' expectations to determine whether or not they are aligned. Because things usually do not happen as quickly as expected, managing frustration is fundamental.

At this stage, it is essential to establish the project leader—the person who will coordinate the work with the team, be the visible face of the venture, and who will not rest until this initiative is successful. This figure is called the *champion*, also commonly referred to as the venture's general manager or CEO (Chief Executive Officer). Leaving this unresolved can have serious consequences for the team and make it difficult to secure private investment.

Team members also have different points of view, so it is important to listen, accept feedback, and learn from mistakes. This also applies to people external to the venture, such as potential investors, evaluators, and advisors, whose contributions can be of great value.

Business Development

A key aspect when forming a spin-off is to have a clear commercial strategy. Although this is a theoretical exercise, it allows for the analysis of the different facets of the business and can become the roadmap for the entrepreneurial team.

In this process, the business opportunity, value proposition, and technical-commercial analysis must be examined. Additionally, the financial analysis should be considered, which facilitates the search for resources. Generally, these analyses are done for the next three to five years.

Some studies have shown that ventures that develop in constant contact with the market, and with participation from their potential customers (e.g., in product design) have a higher probability of success. This is because it allows for continuous validation and potential adaptation of developments through a fluid dialogue between science and the market.

Scientific development toward solving market problems has been a constant challenge for universities and research centers. The promotion of research often does not align well with the actual needs of society and industry. This misalignment can be attributed to insufficient dialogue between researchers and industry stakeholders and their differing perspectives on how to solve the same issues.

It is important to have the business model clear from an early stage, to properly understand the value proposition, and to review it as much as possible with key stakeholders. It is also essential to avoid overestimating the market opportunity or underestimating competitors. Finally, it is vital to carefully choose investors based on the needs at each specific stage and to be specific about what is required from them: strategic support or just money. It is true that in most opportunities, the investors choose the entrepreneurs, so it is essential to build a coherent narrative as a company.

Legal structure is the preliminary step to starting the company, and it is related to managing the legal and administrative procedures necessary to establish the company officially. In some cases, it is advisable to wait until private capital has been raised before incorporating, as investors may set certain conditions for investing. However, if the company can be start operating without raising capital, it can incorporated beforehand.

In the case of technology, it is important to start with an adequate maturity level or viable scaling strategy; that is, creating a venture in the very early stages of technological development can be risky, as financial capital is required to continue its development. It is often easier to obtain public funding for R&D from universities and research centers than from private companies.

It is also important to have licensing agreements in place before establishing the venture so that the technology passes to the venture once it is established. This makes it more suitable in the eyes of a potential investor.

Board of Directors

When formalizing the creation of a venture, it is necessary to establish a board of directors; at this stage, there will be two working teams: the entrepreneurial team and the board of directors or shareholders. It is common for the first boards of directors to be composed of investors, founding team members, and members of the university or research center, if applicable, who are referred to in the literature as "privileged witnesses" since they have been advising the development of the venture.

In one of the few empirical studies in this area, Clarysse et al. (2007) examined the aspects that determine the composition of the board and highlight the tensions that exist between the founding team and other stakeholders (e.g., investors) to determine the composition of the board.

The study shows that, in the case of ventures with external investors with high influence, it is more likely that boards will be developed with skills complementary to those of the founding team; while more autonomous teams tend to form a board with similar capabilities to those already existing (Bjørnåli & Gulbrandsen, 2010).

The study *Dynamics of Science-based Entrepreneurship* (Colombo et al., 2010) suggests that the formation of the board is based on the founders' contact networks. Although the composition of the board may change depending on the stages of the venture's development, its members will always be related to the active members of the venture.

The authors argue that it is essential to consider that at different stages of the venture, the role of the board will change. For example, in the beginning, when those managing the venture have a significant shareholding, the board's exhaustive technical supervision and monitoring lose relevance compared to governance-oriented supervision, monitoring the business model, and support in resource-seeking.

The study goes on to say that as the venture progresses, it is also necessary to professionalize corporate governance. Generally, the role and contributions of the venture's founding teams decrease, and there is a need to institutionalize processes and different sources of knowledge capable of sustaining the venture; the development of an effective board of directors can be important to fulfill these roles.

One of the main functions of a board of directors is to protect shareholder value by ensuring or monitoring the venture, so their composition should consider that external directors contribute new skills and capabilities. In this sense, there are studies that indicate the relevance of the board of directors and the correlation between ineffective boards and the limitations of the company in creating value (Colombo et al., 2010).

Another main function of the board is its strategic role. It is a board's job is to ensure venture's short-term functioning, and progress according to plan. Additionally, it must define the venture's course, detect threats along the way, and often support the general manager through committees on which some of the board members serve.

Shareholders' Agreement

If the decision is made to create a science-based venture, a partnership in which partners are liable only up to the amount of their contributions is recommended. It is also recommended that unanimity of the partners in order to make modifications, such as the bylaws of the company or even the sale of rights, should not be required.

It is advisable to generate the bylaws, which are the equivalent of operating regulations for the venture, immediately upon incorporation. The bylaws are very important and contain essential information, such as the name, capital, and registered office, as well as establishing rights and duties, such as operating conditions.

The following are some guidelines applicable worldwide (ACAFI, n.d.), although there may be some variations depending on the laws of each country.

One of the key aspects of the bylaws, as stated, is the preferential subscription for all shareholders, which is a right that allows former shareholders of a company the possibility of subscribing new shares at the time of a capital increase in proportion to the value of the number of shares they hold in the company.

They also recommend that the bylaws establish the allocation of a percentage of the shares issued in a capital increase to employee compensation plans.

It is important to define how the company will be managed. For example, the number of directors that will make up the board should be indicated; an odd number of members is recommended. If this is not possible, the chairman of the board will decide in case of a tie. It is also relevant to establish who will appoint the members, which is usually done in proportion to the shareholding they have in the company. It is recommended to have at least one external member not related to the partner in order to have a neutral perspective. The minimum quorum for meetings, the frequency of sessions, and operational issues, such as mechanisms for calling, among others, must also be established.

Regarding shareholders' meetings, it is necessary to define the way of convening, the quorums, and the majorities for decision-making in these meetings. Normally, all agreements of the shareholders' meetings are adopted with the favorable vote of the absolute majority of the shares with voting rights, present or represented.

In some cases, it is important to regulate the administration of *subsidiaries.* A subsidiary is a company that is controlled by another. If this occurs, the subsidiary must take the necessary measures to ensure compliance with the obligations of the controlling group's shareholders' agreement.

A very relevant point for shareholders, as stated in the document, is the mechanism for selling or transferring shares or share options. To transfer shares, the consent of the other partners is required. It is very common to incorporate a clause called *right of first refusal,* which means that if a shareholder wishes to transfer their shares, they must first offer them to the company and, if it does not want to buy all or part of them, they must offer them to the majority shareholders in proportion to their participation. If none of them want to acquire them, the shareholder can freely offer them to third parties, but on the same terms offered to the company and the majority shareholders. If there is a third party interested, the shareholder who wishes to sell must inform the company and the majority shareholders.

In the event of a third-party offer to buy shares, the company will have the preferential right to sell, and then the majority shareholders proportionally to their participation. Operational aspects of the transaction, deadlines, and notification mechanisms, among other items, are also regulated here.

There is also the *tag along right.* In the event that a shareholder receives a share purchase offer, even if they are not willing to sell, they must communicate the offer to the majority shareholders, who may sell their shares proportional to their participation in the company. If the majority shareholders are not interested, it will be understood that they authorize the operation.

The forced sale right or *drag along right* indicates that partners can be forced to transfer all shares without price limitation, as long as the bidder is not a related company or has any type of link, and that it ensures shareholders a minimum return of some percentage of their investment, such as three times the investment, which may vary depending on whether they are preferred or ordinary shares.

Ordinary shares are those that give the shareholder common rights as a partner, while preferred shares have certain privileges (e.g., a higher priority in the collection of dividends or in the distribution of the remaining assets in the event of the company's bankruptcy). All these specifications must be included in the shareholders' agreement.

In some cases, it is regulated that shares cannot be sold to a company in which a competitor participates directly, even if it is a minority, unless approved by the board of directors. It is also established that majority shareholders may acquire shares proportional to their participation in new investment rounds of the company. In any case, the new shareholder must assume the conditions of the shareholders' agreement before buying the shares.

It is also very common to establish non-compete clauses; that is, founders are prohibited from participating in new investments, businesses, projects, or productive or commercial activities related to their own businesses that represent competition, or cause harm to the normal development of the company's business, either directly or indirectly. These clauses expire up to a time after the founder ceases to be a shareholder. A reasonable time is two years.

Information clauses are also established to keep shareholders informed, as well as non-compliance clauses, which indicate the validity and reasons for the termination of the agreement.

Procedures for communications and notifications to partners are established in the general provisions, including what happens in the event of a company restructuring that involves a change to a foreign jurisdiction. In this case, it is established that the parties will make their best effort to maintain preferences such as those existing in the current agreement. An arbitration mechanism must also be included in the event of any controversy between the partners, and the entity which will carry out said arbitration must be defined. In addition, the law governing the agreement must be indicated.

Finally, it is recommended that the following items should also be included: annexes with information about who the partners are with their respective percentages, which is called a *cap table*; definitions of terms used in the agreement are also included so that they are not open to interpretation; an adherence form to the agreement, which is necessary when new partners enter; and, an item detailing the fundamental matters and quorum for making certain agreements.

Establishing the shareholders' agreement in a venture is essential, as it will allow making fundamental agreements, such as the distribution of profits. It will also make it possible to protect minority shareholders from any shareholder and the shareholders of a company from a new majority partner, giving them the possibility, for example, of selling their shares at a fair market price. In short, it seeks to regulate the interaction between partners and increase the chances of success for all parties.

Importance of the "*Exit*"

A fundamental element for investors that is often not addressed by science-based entrepreneurs is the *exit*, which refers to the way the investor will cease to be a partner and make a profit along the way; in other words, it is the way to monetize investments in ventures, seeking to generate attractive financial returns.

When starting the capital raising process, in addition to having a clear value proposition, knowing who the customers are, and the state of technology development (among other aspects), it is important to clearly show the "exit projection," i.e., the moment when the investor can exit the invested company, and materialize the profitability generated by their investment.

In this type of investment, it is very common for the investor to deeply analyze the probabilities of materializing a good exit—establishing the moment in time, the moment of the company, and the exit conditions.

Vicente Esteve analyzes how curious it may seem that someone who has just founded their own company, perhaps not yet achieving sales and realizing that it will be very complex to get it started, is thinking about selling it. This begins to make sense when analyzing the definition of science-based entrepreneurship, which, as Steve Blank proposes, "is a temporary organization that seeks a repeatable, profitable, and scalable business model" (Esteve, 2014).

In this sense, it is essential to know the best moment for an exit. Unfortunately, or perhaps fortunately, there is no single moment or best moment, as this will depend on multiple factors and, above all, on the profile and expectations of the investor partner. However, there are different moments that will allow making the exit decision. For example, when there is the possibility of selling the company to another company, which is usually in the same industry and generates synergy with the venture, or also with the entry of a majority investor.

Theoretically, the best moment for an exit is when the market value of the company is higher, which occurs when the product is in the ascending phase of the initial sales stage. However, and especially in science-based ventures, the exit can be much earlier, even in product development phases.

When suggesting an exit plan to investors, it is also important to consider the potential gains that this may imply, which is usually measured based on some multiple of the related investment. Tax issues must also be considered when deciding the type of company to create the type of potential investors. It must also consider where the funds may come from, which, having been well structured in their creation, make the exit much easier. In science-based ventures, the exit should be from three or four times, up to several tens of times, the amount invested.

Enabling Conditions for the Creation of Science-Based Entrepreneurship

Entrepreneurial Capacities

There is no single path to successfully developing a science-based venture; among many other factors, it is greatly influenced by the context, the type of technology, and market evolution. However, there is considerable consensus regarding a couple of key factors that make a difference.

Bonardo states in the document *The M&A Dynamics of European science-based entrepreneurial firms* (Bonardo et al., 2010) that science-based ventures require the availability of two critical resources: human capital, for the creation and management of innovations; and financial capital, for investment and growth.

In the publication *Effectuation and causation in science-based new venture creation: A configurational approach* (Villani et al., 2018), it is argued that the capacities of the entrepreneurial team are key to increase the probability of success of a science-based venture. Those ventures that have, among the capacities of their researchers, business experience and commercial knowledge, risk-taking propensity, contact networks, and flexibility towards challenges and environmental changes have a higher probability of success.

Another study lists the different factors that are key to success in science-based ventures: proximity to world-class research centers; access to highly qualified personnel; and funding from various sources to support the different phases of the venture. In some countries, quality of life factors (e.g., schools, community, and services) play a role. Another important factor is having appropriate, adaptable, and affordable laboratories, as well as office spaces. The business environment that supports and encourages entrepreneurs to start and grow ventures, the availability of support service providers to create an entrepreneurial ecosystem, access to attractive markets, and a favorable government environment with financial and tax incentives also influence success (Ahn & Meeks, 2008).

Although one of the key aspects for the creation and success of a science-based venture is the entrepreneurial capacity of its founders, paradoxically, in most countries there is a lack of training opportunities to guide and prepare scientists, and direct them towards the path of academic entrepreneurship.

Institutional Support

In the study *Science-based technological ventures in Latin America* by the Inter-American Development Bank (Kantis & Angelelli, 2020), it is argued that "universities offer a strategic context for this type of ventures, not only because of the potential that arises from the results of research projects, but also because of the knowledge that circulates more broadly through professors, graduates, and master's, doctoral, and even undergraduate students in the various activities that take place in higher education institutions."

In the publication *Dynamics of Science-based entrepreneurship*, another study (Colombo et al., 2010) analyzed the role that universities play in the development of science-based ventures. They conducted an empirical analysis and concluded that proximity to geographical areas with universities contributes to the growth of technological ventures. It is important to note that the scientific quality of the research carried out by universities has a positive effect on the growth rates of science-based ventures, although the commercial orientation of the research has a negative effect.

The study *The M&A dynamics of European science-based entrepreneurial firms* (Bonardo et al., 2010) suggests that creating more direct links between universities and the business world represents an effective solution to promote the creation of science-based companies. Ventures linked to a university have beneficial effects on reputation, which in turn facilitate collaboration with other organizations and improve the credibility of the project in the eyes of third parties, such as companies or investors.

The study concludes that science-based ventures have a higher probability of being acquired by another company. This is very relevant in these types of ventures, as they are characterized by having low operational efficiency, and a different business vision than other types of ventures. Being acquired by others allows the establishment of new ownership and management structures in order to improve their productivity and maximize financial return. Also, these types of ventures are attractive due to their innovative capacity; being linked to a university could imply a certification role, improving their appeal in the eyes of other companies.

However, organizations that conduct research and development have few incentives or stimuli (such as clear rules and frameworks for action, incentives in academic and professional career development, monetary incentives, and recognition of achievements, among others) for their academics to become interested in entrepreneurship.

Another factor that affects the creation of science-based ventures is related to the inexperience of the offices or departments responsible for promoting such activity within institutions. This is compounded by the lack of articulation with the ecosystem, such as other universities or the public sector, but above all, the business sector.

A factor that is increasingly gaining strength when developing science-based ventures is the regulation of the conflict of interest of the researcher who is linked to an institution. Conflict of interest is understood as the opportunity that the academic has to exert influence on the decisions made by the institution, which may imply personal gain or that the academic has direct or indirect interests that affect their independent decision-making in the fulfillment of their duties with the university.

Public Policies

In the publication *Dynamics of Science-based entrepreneurship* (Colombo et al., 2010), the arguments for why a country should support the development of science-based ventures are analyzed. First, it argues that there is an information asymmetry problem, which means that investors may be exposed to *moral hazard*. This refers to a situation in which one party engages in risky activity, knowing that it is protected against the risk and that the other party incurs the cost, which arises when both parties have incomplete information about each other. This is one of the reasons why there is a low incentive to invest, and government intervention is necessary, especially in the early stages.

On the other hand, when there are radically innovative technological developments, sometimes science-based ventures generate social benefits much greater than the private benefits that investors can capture. On some occasions, it is difficult for these companies to effectively protect their developments from competitors, which reduces the possibilities of private investment.

Another inhibiting factor of early investment is the level of risk to which science-based ventures are subjected. "The risk associated with the creation of innovative companies may be too significant to be assumed by a private actor. There are uncertainties related to technology, uncertainties about the existence of sufficient demand, significant fixed costs that are not easily recoverable, etc. The intervention of public authorities allows these companies to finance their risky projects at lower costs" (Colombo et al., 2010). However, it is important to consider that government support could generate distortion in the reality of the ventures, artificially reducing the risks, and could "subsidize inefficiencies and provoke competition distortions" (Santarelli & Vivarelli, 2007).

Another way the government can contribute to the development of science-based ventures is by becoming a customer of them. This is done through policies and programs that stimulate the demand for goods and services resulting from technological innovations, such as "innovative public procurement" or "challenge-based initiatives". In addition, successful international experiences indicate that the demands generated by an entrepreneurial government, such as through spending on defense, health, aeronautics, and others, can drive dynamic processes of innovation and entrepreneurship (Mazzucarto, 2014).

There are multiple studies that show that science-based ventures are highly risky, and a very small portion, close to 2%, grow rapidly and even become listed on the stock market. Because of this, it is a profound mistake of public policy to measure the success of a science-based venture by the number of jobs created or financial returns and to condition support for the development of science and technology-based companies on this. Governments should contribute to the promotion of science-based ventures, as they contribute to the well-being of society and enhance the competitiveness of countries through the development of better goods and services.

Financing

Science-based ventures generally require a large investment in the early stages, so it is crucial that there is an appropriate financing chain that runs from proof of concept to productive scaling; this lengthy period makes the participation of risk investors highly necessary. There are few studies that have analyzed the relationship between the participation of a risk investor and the success of a company. While some (Audretsch & Lehmann, 2004; Engel & Keilbach, 2007; Hellmann & Puri, 2000; Manigart & Hyfte, 1999) have shown that scientific ventures with risk investment reach the market faster and with greater growth, other studies (Bottazzi & Da Rin, 2002; Murray & Burgel, 2000) conducted in Germany and the United Kingdom, concluded that risk investment has no impact on company growth.

There are few studies on the relationship between venture capital investments and the capabilities of the entrepreneurial team. One of them (Colombo & Grilli, 2010) analyzes the influence of the entrepreneurial team on the growth of high-tech ventures that have been supported by risk investors. Using a sample of 439 Italian companies, it was discovered that once a company receives risk investment, the role of the founders' skills becomes less important in contributing to the growth of the company. This result points to the importance of the function of *coaching* fulfilled by venture capital investors to improve the performance of the portfolio of companies.

Due to the above, it is very necessary that the initial stages of ventures be financed with government support, but many times, the existing financing is more oriented towards ventures and less intensive in research and development. The succeeding stages of the financing chain are generally unfunded, such as prototyping and packaging, and at the final stage of the chain, investors appear who are looking for robust projects. However, if there is a financing gap in part of the development of the ventures, it is very natural that the projects involve a lot of risk or simply do not reach the stage of private financing.

In conclusion, the most relevant aspects when undertaking a science-based venture are the capacity and quality of the research and development carried out by researchers and their respective institutions as well as the proximity and interaction with the business sector. Factors such as the size and dynamism of the market affect the possibility of amortizing and making profitable investments in R&D (Van Roy & Nepelski, 2017), while the sophistication of the demanders affects the degree of acceptance of innovative proposals.

This type of venture is capital-intensive, so it requires a good financing chain throughout the process. It is also crucial to consider all regulatory aspects in order to carry out the venture.

As has been proposed, one of the fundamental factors is the entrepreneurs themselves, so generating the appropriate incentives and instances of training, education, and complementing capacities is central to increasing the chances of success of these businesses.

As will be seen in the following chapter, the creation of science-based ventures requires a combination of different financing sources, depending on the different stages of the venture's life cycle. In some cases, they are self-sourced or both public and private sources.

It is very important to carry out planning that establishes the different stages in which capital, amounts, and anticipated origin is required. However, it must be flexible planning and adapt to reality and changes in the environment.

In this sense, the lack of financing is a perilous constraint for ventures in general and, in particular, for science-based ventures, as they usually demand long development periods and significant resources before the expected results are achieved.

References

ACAFI. (n.d.). *Shareholders' agreement.*

Ahn, M. J., & Meeks, M. (2008). Building a conducive environment for life science-based entrepreneurship and industry clusters. *Journal of Commercial Biotechnology, 14*(1), 20–30. https://doi.org/10.1057/palgrave.jcb.3050076

Audretsch, D. B., & Lehmann, E. E. (2004). Financing high-tech growth: The role of banks and venture capitalists. *Schmalenbach Business Review, 56*(4), 340–357. https://doi.org/10.1007/BF03396700

Bjørnåli, E. S., & Gulbrandsen, M. (2010). Exploring board formation and evolution of board composition in academic spin-offs. *Journal of Technology Transfer, 35*(1), 92–112. https://doi.org/10.1007/s10961-009-9115-5

Bonardo, D., Paleari, S., & Vismara, S. (2010). The M&A dynamics of European science-based entrepreneurial firms. *The Journal of Technology Transfer, 35*(1), 141–180. https://doi.org/10.1007/s10961-009-9109-3

Bottazzi, L., & Da Rin, M. (2002). Venture capital in Europe and the financing of innovative companies. *Economic Policy, 17*(34), 229–270. https://doi.org/10.1111/1468-0327.00088

Clarysse, B., Wright, M., Lockett, A., Mustar, P., & Knockaert, M. (2007). Academic spin-offs, formal technology transfer and capital raising. *Industrial and Corporate Change, 16*(4), 609–640. https://doi.org/10.1093/icc/dtm019

Colombo, M. G., & Grilli, L. (2010). On growth drivers of high-tech start-ups: Exploring the role of founders' human capital and venture capital. *Journal of Business Venturing, 25*(6), 610–626. https://doi.org/10.1016/j.jbusvent.2009.01.005

Colombo, M. G., Mustar, P., & Wright, M. (2010). Dynamics of science-based entrepreneurship. *The Journal of Technology Transfer, 35*(1), 1–15. https://doi.org/10.1007/s10961-009-9114-6

Engel, D., & Keilbach, M. (2007). Firm-level implications of early stage venture capital investment—An empirical investigation. *Journal of Empirical Finance, 14*(2), 150–167. https://doi.org/10.1016/j.jempfin.2006.03.004

Esteve, V. (2014). *Exit in startups.* https://vicentesteve.com/planning-the-exit-in-my-startup/

Hellmann, T., & Puri, M. (2000). The interaction between product market and financing strategy: The role of venture capital. *Review of Financial Studies, 13*(4), 959–984. https://doi.org/10.1093/rfs/13.4.959

Kantis, H., & Angelelli, P. (2020). Science and technology-based ventures in Latin America: Importance, challenges and recommendations for the future. In *Science and technology-based ventures in Latin America: Importance, challenges and recommendations for the future.* Inter-American Development Bank. https://doi.org/10.18235/0002156

Lubik, S., & Garnsey, E. (2016). Early business model evolution in science-based ventures: The case of advanced materials. *Long Range Planning, 49*(3), 393–408. https://doi.org/10.1016/j.lrp.2015.03.001

Manigart, S., & Van Hyfte, W. (1999). *Post-investment evolution of Belgian venture capital backed companies: An empirical study.*

Mazzucarto, M. (2014). *The entrepreneurial state: Myths of the public sector versus the private sector.*

Menzies, M. B. (2012). Researching scientific entrepreneurship in New Zealand. *Science and Public Policy, 39*(1), 39–59. https://doi.org/10.3152/030234212X13214603531842

Murray, G., & Burgel, O. (2000). The international market entry choices of start-up companies in high-technology industries. *Journal of International Marketing, 8*(2), 33–62.

Santarelli, E., & Vivarelli, M. (2007). Entrepreneurship and the process of firms' entry, survival and growth. *Industrial and Corporate Change, 16*(3), 455–488. https://doi.org/10.1093/icc/dtm010

Sethi, A. (2016). From science to startup. In *From science to startup: The inside track of technology entrepreneurship*. Springer International Publishing. https://doi.org/10.1007/978-3-319-30424-3

Van Roy, V., & Nepelski, D. (2017). Determinants of high-tech entrepreneurship in Europe. *Business*. https://doi.org/10.2791/96153

Villani, E., Linder, C., & Grimaldi, R. (2018). Effectuation and causation in science-based new venture creation: A configurational approach. *Journal of Business Research, 83*, 173–185. https://doi.org/10.1016/j.jbusres.2017.10.041

Financing

6

Abstract

This chapter addresses the issue of financing in the technology transfer process, from research and development to the consolidation of technology in industrial applications. The different stages of financing and the mechanisms used in each one are described. It includes details of the different stages of an entrepreneurial venture and the required financing according to the stage. In addition to describing financing mechanisms, a variant of the traditional "3F" model is proposed with the inclusion of "Smart," highlighting the importance of trust and intelligence in early investments. Investment modalities in entrepreneurial ventures, crowdfunding, venture capital, corporate venture capital, and private equity are also discussed. Financing in technology transfer is a challenge that requires different sources of financing at each stage of the process. The government plays a fundamental role in the early stages, while angel investors and other private actors contribute capital and knowledge in later stages. The mentioned investment modalities play a crucial role in the entrepreneurial ecosystem, providing financing, support, and growth opportunities.

Keywords

Financing • Technology transfer • Government subsidy • Entrepreneurship • Angel investors • Seed capital • Venture capital • Crowdfunding • Corporate venture capital • Private equity

A. Ossa, *From Research to Market*, https://doi.org/10.1007/978-3-031-71340-8_6

Financing Stages

One of the major challenges in the technology transfer process is financing throughout the entire chain, from research to the arrival of its results in the market or society. If the chosen path to transfer a technological development is the creation of an entrepreneurial venture. In that case, it is necessary to go through a long and complex path in search of resources. In this sense, the process contemplates different stages, with various risks and different investment amounts.

The process can be divided into four main stages. The first is research and development ("R&D"). In this stage, the idea is conceptualized, initial experiments are carried out, and proof of concept developments are made up to laboratory validation. Traditionally, this phase is heavily financed by government subsidies. A subsidy is understood as non-repayable financial aid, generally provided by the government, considering that it can promote the development of knowledge and economic growth in areas where the private sector does not participate due to high technical and commercial risks.

The second stage is the gestation stage. This is when real-scale pilots are developed, and validations are carried out in a real environment. Private financing, known as the 3F's (friends, family, and fools), begins to emerge; when including the entrepreneur's own funds (founder), it is also known as the 4F. Angel investors may appear here, consisting of early-stage risk investment, assuming high risk and betting on a high return. Unlike the 4F, who often invest without analyzing the return, angel investors thoroughly analyze this variable before investing. Finally, there are seed capital funds (also government subsidies), which help boost developments that still present considerable risks.

The third stage is the growth stage. Here, prototype demonstrations are carried out in a real environment, and the first commercial-scale pilot is carried out. Private financing predominates through venture capital investment. Some reinvestment of revenues from sales made, if any, is also generated, and crowdfunding begins to appear. To a much lesser extent, some government subsidies may also be obtained at this stage.

The fourth stage is the consolidation stage. The technology is already in industrial application. The company is already selling and operates with private capital, such as risk investment, bank loans, or traditional investment instruments as well as reinvestment of revenues from sales made.

Financing Mechanisms

The different financing mechanisms and financiers, respond to various motivations. The government is motivated by promoting national development and does not expect a financial return. In the case of private investment, e.g., angel investors or venture capitalists, motivations are related to direct economic return, although also to the economic development of the country (Ahn & Meeks, 2008).

Government Subsidy

The government plays a fundamental role in the innovation process. In some cases, it assumes a subsidiary role, especially in the early stages, and in others, even a guiding role in innovation strategies. Without the role of government, it is very difficult to promote, at least in the early stages, innovation in countries since the risks are very high, both technical and commercial, for the private sector to assume. It is also true that investing in research and development allows for more solid economic growth; it contributes to a more efficient and more productive economy and increases the development of goods and services with higher added value.

A subsidy is an economic contribution that does not have a consideration and is non-reimbursable. Generally, these subsidies are provided by the government because of the reasons expressed above, related to the volume of risks and the interest government has in promoting development.

The characteristics of the subsidy (amounts, terms, conditions for awarding them, etc.) depend on the state of development of the technology. In earlier and less appropriable stages, a larger subsidy with lower entry conditions and sometimes with longer terms, compared to future subsidies, is assumed. Generally, in these stages, the goal is to reduce technical risks. In later phases, as some of those risks are controlled and a certain state of technological development is met, the demands become greater. In some cases, a financial counterpart from companies is required.

In the R&D stage, a strong government subsidy is required, and, in some cases, it not require a business counterpart. In this stage, technical risks are expected to be reduced. In deciding whether or not to assign a subsidy, the most influential factor is the methodological coherence of the proposal. The participation and contribution of the private sector always help get the proposal a better evaluation, since it is an indicator of the relevance of technological development.

In the next stage, a business counterpart that makes financial contributions is required. When assigning funds, some technical risks are expected to be controlled, and technical and commercial risks are evaluated. In this stage, the research and management team that carry out the project and the institution or company under which the research is being developed begin to take on relevance.

In the following stages, the subsidy becomes less relevant compared to the contribution of private individuals. Only in some cases are state subsidies maintained, such as in the case of seed capital, which is necessary to launch the technology because the risk is still high at this stage.

From the 4F to the 3F + S

For several decades, the term 3F has been used to explain where to obtain the first non-state financial resources to continue the development of technology, referring to friends, family, and fools. After a while, a fourth F was incorporated, referring to the founders, who invest part of their capital at this stage, so the model is known as the "4F."

Here a variant of the theory is proposed; instead of talking about fools or crazy people or lack of understanding, we will talk about "Smart." That is, we will talk about founders, friends, family, and Smart (3F + S Model) since over time, it has been shown that fools or crazy people are rather intelligent people who have made a bet in very early stages.

As Silvia de Torres Carbonell states in her text *Financing sources for innovators in Argentina: venture capital, angel investors, institutional investors* (Carbonell, 2012), the first to invest in these ventures are those whom the entrepreneur manages to convince to accompany them on this entrepreneurial adventure. That is, instead of doing it out of love or foolishness, as the 4F model suggests, they do it out of trust or intelligence.

In these cases, the capital raised is relatively small because the project is in the idea stage or at the conceptual stage. It serves to conduct research, face initial operating costs, and prepare the venture to present it to angel investors, who are the next link in the financing chain, as proposed by De Torres. Although, at this stage, the money invested is small, the multiplier effect it can generate is very significant. Vicente Esteves indicates that the investment at this phase has a key social impact on the development of the entrepreneurial ecosystem. However, those who provide it are likely the "great forgotten," at least in terms of recognition, when the venture is successful. "It has a certain romanticism that distances it from strictly financial investment and brings it closer to other actions that leave a social footprint" (Steve, 2014).

For the system to work, the investor must not need to get back the money contributed, neither in the short nor in the long term; that is, they must be very willing to lose the entire investment. With this, the entrepreneur must focus on survival, because, although it may be little money, it may be one of the most important investments in the history of the venture as it corresponds to the contribution of the first ones to believe in the project. Finally, there must be good communication between the entrepreneur and the investor so that the entrepreneur constantly informs the investor of the progress made with the investment and possible liquidity or profitability problems that may arise (Steve, 2017).

Since the investment is made primarily for personal relationships rather than interest in financial return, loans are typically interest-free with no deadlines for repayment. In general, it is easier to convince this group than investors in later stages.

Angel Investors

Angel investors are a very relevant link in the financing chain of science-based ventures. In this sense, Carbonell states:

> Angel investors are individuals with a certain net worth who offer venture capital to entrepreneurs with whom they have no family or friendship ties. They play a key role in

> the development of entrepreneurial activity, both in terms of financial capital and their contribution in terms of skills, experience, and contacts acquired in their professional life. (Carbonell, 2012)

Additionally, they enjoy accompanying a technological venture due to the risk involved, which usually reminds them of their own venture.

This type of investor contributes not only money but also knowledge about the market, business management, their contact networks, and other relevant areas to support ventures in their initial stage, which is called "smart money."

At this stage of development, because the technical and commercial risks are still very high, investors expect a high financial return (even five to ten times the amount invested within five to seven years), although their main motivation is to contribute to the development of the venture. While it depends on the reality of each country, investments usually range between $500,000 and $1 million. In general, they are patient investors who wait and accompany the entrepreneur in their difficult process; they do not take a majority stake in the ventures and look for projects with high growth potential (Carbonell, 2012).

If the investment is so risky, Carbonell continues, if there is uncertainty about the term of the investment return, if—although the return is high—there are few chances of achieving it, one may wonder why angel investors are willing to participate. The answer is that because they are people who have economic stability and are generally familiar with the world of risk investment in their businesses, they have the desire to "give back" to society and to give others the opportunities they had. On the other hand, it is suggested that they are driven by the adrenaline of being part of an emerging company without the pressure or obligation to dedicate themselves to it or assume its executive responsibility. Of course, there is a business or investment model behind every angel investor; since they expect the return on investment to be very high (between seven and ten times the amount invested), they also expect that this return covers the losses that the rest of the other projects they invest in surely bring, as statistics show that nine out of ten investments fail. Therefore, an investment portfolio is usually created, atomizing their investment into several projects.

Over time, often driven by government or even self-managed, is the formation of networks or clubs of angel investors has been happening. These networks facilitate the exchange of information and encourage investment, even when larger amounts are required in technologies with greater uncertainty.

Family Offices

Another investment route is a "family office," which is an investment platform for people with a lot of money who seek to have professional management of the wealth they have. They are called family offices because they manage the wealth of a single family, and their main focus is to manage that capital. The goal is the transfer of wealth between generations and, since transferring wealth to new generations can be very costly at the tax level, and since the next generations may not be interested in managing it directly or could gradually destroy it if they do not know how to manage it, these entities are created. Generally, those in charge of these funds are new generations of the family, that is, the children or grandchildren of those who contributed most to the creation of that wealth. However, it is important that whoever manages the fund is independent of the family, to achieve the degree of objectivity and professionalism necessary in these cases (Sethi, 2016).

The pioneers in these matters were the Rockefeller and Morgan families in the United States, who created specialized areas precisely to professionally manage their wealth.

Although it is true that family offices are created by a single family and are dedicated exclusively to the wealth of that family, there are also *multi-family offices*, i.e., offices that manage the wealth of several families. They usually start as single-family offices and over time, to generate more asset volume and achieve economies of scale, they become multi-family offices.

In the case of these platforms, the investment criteria tend to be broader than those of purely financial investors, so this is where startups can take advantage of opportunities to obtain investment. Individual investments can be relatively large depending on the size of the fund. In addition, the time horizon of investments tends to be much longer than that of financial investors and can exceed five or even ten years. To access a family office, references are essential, as trust must be gained for them to invest (Sethi, 2016).

The family offices have another advantage: even if the economy is bad, they can still invest more and in larger rounds.

Crowdfunding

"Crowdfunding" is a trend that fills existing financing gaps for entrepreneurs. The system consists of people providing money to the venture before the development of the respective technology, which allows for the production of goods or services that are then delivered to these same people.

This modality is gaining more and more strength, as it is a solution to the obstacle faced by ventures with good product ideas that simply cannot obtain financing from more conventional investors.

The benefit of crowdfunding is that the venture does not dilute—it does not give up equity in exchange for the money raised; only the committed products are delivered to the individual contributors.

The customer-investors act as "early adopters," who take on some risk by purchasing new and untested products. However, the amounts they invest are low, and they can feel like they are investing or helping entrepreneurs move forward. Additionally, these early adopters long to access technology before others.

Crowdfunding presents various advantages. It allows financing to be raised at a complex stage of development and for small amounts, although at this stage it is usually easier to raise large funds than small ones. It also helps to have a customer portfolio, which is highly valued by investors. Finally, it allows products to be tested in the market and feedback to be obtained from end users, thus estimating which attributes of the product are of the highest perceived value for end users. It is well known that many new products are used in ways different from what the founders originally envisioned.

The risk of obtaining financing through crowdfunding is that, if the products do not meet the expectations of the customers, a negative impact on the reputation of the venture could be recorded at a time when the company cannot afford it (Sethi, 2016).

Crowdfunding works through companies that specialize in raising investment. Entrepreneurs send their projects to one of these platforms, which usually operates online, for evaluation and acceptance. They are asked to send a description of the project, how much money is required, and how much time there is to raise it. The initiative is then published, sometimes a dissemination campaign is launched by the crowdfunding platform. At the end of the period, the amount raised is verified. If the capital is raised, a percentage is paid to the platform, which generally fluctuates between 2 and 7%. However, some have a fixed payment system or a combination of fixed and variable payments.

As in most of the different stages of investment, it is strongly influenced by the cultural environment. While risk investment is a daily activity in the United States, investors in Latin America are much more cautious. This is quite similar in the case of crowdfunding, and these "micro" investors are also less prone to risk in Latin American countries.

Venture Capital

"Venture capital" funds are those that invest in companies in the early stages of formation, with high risks in the technical and, above all, in the commercial aspects.

Because venture capital invests in high-risk ventures with high potential returns, their diversification strategy is very important. Therefore, these risk investments are made both in early-stage ventures that have been recently created and do not yet have sales and in growth-stage ventures that have already managed to raise a small investment and have had the first sales. They generally invest in a large number of companies and thus diversify their investment. Although some of them may fail, the bet is that another yield enough to cover the losses of the entire portfolio and, even more, generate great returns. This implies that venture capital funds generally invest through minority ownership stakes, which is why it is common to find several investors in a single venture. The ventures they invest in are in the formation stage or have been started between three and five years ago. Some funds called "early stage" could invest in ventures that do not yet have sales or have only recently started their commercialization process. Generally, these types of ventures are intensive in intellectual property and develop cutting-edge technologies or are based on innovative business models. However, the most valuable part of the company is its entrepreneurs. In fact, venture capital funds decide on investment based on the capacity of the entrepreneurial team rather than the potential of the technology itself.

It is well known that science-based ventures do not fail because the technology is bad or because the entrepreneurial team is not suitable; in general, they fail due to a lack of financing to carry out product development and bring the venture to market. In this sense, the venture capital industry comes to fill part of the traditional financing gap faced by ventures. Usually, the investment amount range between $300 K and $8 M and are aimed at initiatives that face the impossibility of financing through traditional channels.

Venture investors put their money in early-stage or nascent ventures in exchange for an equity stake in the company, with the expectation of obtaining a high return on their investment—four to ten times the amount invested, or on the order of 35% to 50% annually, if the investment decision was correct.

If the venture is successful, it is in a position to be sold to another company, obtain financing from new sources, or go public; however, studies indicate that this last path only occurs in 1% or 2% of cases. When one of these alternatives occurs, the investor multiplies their value; they then have the option to exit and obtain liquidity or stay and bet on continuing to increase their value. It is estimated that only one or two out of every ten investments are successful. However, the success is expected to be such that it compensates for the losses suffered in the rest of the investment portfolio.

Usually, venture capital investments are channeled through "venture capital funds," which obtain capital mainly from institutional investors, individuals, and even family offices or university funds; in some countries, pension funds, insurers, or other financial entities are allowed to invest in these funds. Once formed, individual investors do not participate in the management of the fund. This is in the hands of a manager, who is responsible for managing the companies in which they invest, thus facing great contingencies. Venture capital funds usually last eight to ten years. In the first half of the fund's life, investments are made according to its investment policy; in the second half, the "exit" of investments is structured so that the fund can recover the invested capital and realize the return of those investments that have been successful (Carbonell, 2012).

Corporate Venture Capital

The concept of "corporate venture capital" corresponds to an investment capital fund of a company that is focused on investing in ventures.

Large companies began to create venture capital funds, as they generally had problems developing truly disruptive innovation and only generated operational innovations or business models. Along the way, they realized they needed ventures to support them in developing innovation to revolutionize the market.

Corporate venture capital invests in ventures for a variety of reasons: to solve strategic problems of the company itself; to find novel goods or services to sell or resell; or, to find new technologies to replace existing ones. Another reason may be to reach new markets that are very complex to achieve with current capabilities or to enhance existing business units. In this sense, corporations seek a strategic benefit in addition to the economic benefit inherent in a risk investment.

Normally, large companies can create this type of investment fund, and they seek to create structural collaborations with ventures so that both can grow. However, it is not always easy to create corporate venture capital.

When these structures are created within companies, sometimes there is a lack of a clear and tangible value proposition for the corporation. In addition to this, there is often a lack of conviction among the corporation's top executives and a lack of understanding on the part of the corporation's employees to comprehend what this area is doing. As corporate venture capital develops, other complications begin to emerge, such as the lack of a clear roadmap and a lack of financial resources to expand the unit; sometimes the corporation's expectations are not met.

The relationship between the corporate giant and the nascent venture is not always easy because corporations are generally quite rigid, and ventures are more volatile. However, investments of this type allow ventures to begin their takeoff, and companies to integrate and innovate in a way they could not have with the capabilities they have.

Private Equity

"Private equity" is a venture capital fund focused on established companies, where a majority stake in the company is acquired. It is usually temporary and sold after the company has gained more value.

This type of fund does not have a specific orientation when it comes to selecting companies in which to invest. Rather, it invests in all types of companies and various industries without distinction, so it assumes a high risk. It usually invests a significant amount of money, acquiring a majority stake in a company or even up to 100% of the company. This type of investment is made in more advanced stages of a venture's life cycle, and the entrepreneur's role tends to be less prominent.

These funds are also characterized as long-term financing for already established companies that present opportunities for regional expansion or expansion in their lines of business. Generally, when buying majority positions, private equity funds actively participate in the management of the companies in which they invest. The exit of a private equity investment is usually via a strategic investor, another fund that acquires the company, or even through the Stock Exchange (Gompers et al., 2015).

In addition to the significant monetary contribution they make, one of the great contributions made by private equity funds at this stage of the financing process of the science-based venture ecosystem is the generation of support in the daily management of the company. This support helps to generate the necessary credibility to facilitate contacts with third parties, such as banks, suppliers, and customers.

In general, they share the operating mechanism, the philosophy for investing, and the contribution they seek to make to the company that venture capital funds support. However, private equity funds assume a much lower risk in the investment, compared to that of a venture capital, although the amounts to invest are also much larger. For this reason, they usually carry out a smaller number of operations each year since the risk of failure is lower and they do not require a diversification strategy. The expectation is that the return on investment is twice what is expected in a venture capital investment.

Others

In addition to the entities that exist for each of the different stages of financing a venture, there are "non-profit foundations." These are private, legal entities without profit motive, endowed with their own assets, which offer a fund that is usually provided by the founders for personal motivation or sometimes by testamentary mandate. These organizations are created with a specific purpose, legally set forth in their social objective, which must be faithfully pursued. It is important to note that a non-profit organization does not restrict the conduct of commercial or profitable activities in order to increase its assets but only prevents profits from being distributed among its founders and requires that profits be reinvested in the foundation's own activities for the purpose for which the foundation has been created.

Not being directly dependent on the government, nor pursuing the maximization of monetary benefits as their ultimate goal, they have greater freedom to make decisions with the aim of contributing to private social investment and the return to the community itself.

Foundations are, in essence, supportive. Their work is done out of charity, compassion, or their own interest or conviction to help other people or society as a whole. Although it is true that foundations have been created for charitable or cultural purposes, some time ago, organizations emerged that sought to promote science and technology, and they long for research results to impact the market. Under the shelter of large fortunes, foundations have been created to combat, for example, rare diseases, promote scientific development to solve climate change problems, or even support the creation of science-based ventures.

Risk Investment

The entrepreneurial team is the fundamental factor in convincing an investor to invest in the venture. However, other factors influence the decision-making process.

One of the relevant factors is the business opportunity, that is, the degree of relevance of the problem detected, how much impact it has on the industry where it is developed, and the size of the market. The latter implies not only analyzing the total size but also whether or not it is achievable and the time required to reach that market.

Another aspect that is analyzed is the solution proposed by the entrepreneurs. The level of innovation of the proposed idea must be considered: What improvement does it represent compared to the current state of the art? Technical and commercial risks are also assessed, such as the probabilities of achieving the proposed technical solution and obtaining the first customers; and especially in science-based ventures, the intellectual property protection strategy is analyzed.

The level of progress and validation of the development are other fundamental factors that investors analyze when making an investment decision. In this sense, the current state of development related to scientific validation is evaluated, such as: the pilots carried out; whether certifications, patents, or regulations have been obtained; and whether awards or recognitions have been granted. The level of data collected allows for greater certainty of the technical effectiveness of the solution. From the validation point of view, actions such as formalized alliances with third parties or if they have achieved sales or potential interest from other investors, if applicable, are elements that help reduce uncertainty.

The financial plan is part of what is analyzed. Although it is true that this is a projection, every good financial plan must have solid foundations. That is, it must be developed with sensible and well-founded hypotheses; if some of this data is validated, it is even better. However, there will always be a component of assumption.

Finally, a critical factor is the team. The knowledge that its members have about the subject, the commitment, and the execution capacity (such as the dedication time) of the entrepreneurs are analyzed. It is not expected that they possess all the capacities, but whether the team has the ability to learn and collaborate with each other and with the investor is analyzed. Although it is true that investors evaluate the set of capacities, they also assess the level and formality of commitment that the founders have with each other in determining the risk of abandonment by any of them.

A very relevant factor is whether the exit opportunity is clear and attractive: this is known as an "exit.". If things do not go well, a liquidation will be initiated, which is not an *exit* but an *end*.

Although it is true that it is very complex to obtain risk investment, it is important to understand that not every investor is good for the development of the venture. In this sense, there are several types of investors, and it is also true that there is no correct or incorrect investor, per se. In that context, since a partner who contributes to the success of the venture is being sought, it is very relevant to generate trust beyond the simple financial contribution. Therefore, the only correct investor is the one whose interests are aligned with the entrepreneur's vision for the commercialization of the venture and its partners (Sethi, 2016).

To choose a good investor and not just be chosen, one must deeply analyze the investor's profile, e.g., whether they seek to be an active actor in the management of the venture or if they will leave that responsibility to the founders. It is recommended to look for an investor who complements the role that the founders seek to perform. It is also important to analyze whether the new partner or partners have knowledge in the industry or if they have invested in related ventures previously, which may imply that they add great value to the current business. Another factor that can be very interesting is if the investor has been an entrepreneur previously. This is key since they understand the difficult path that must be traveled and could probably be much more empathetic than someone who has not necessarily had this experience.

It is always good to reflect on who can be the companion on this journey—the financial contribution is not the only thing to consider, although it is clearly one of the relevant things. One of the easiest and most cost-efficient ways to get to know someone is to take a trip together. In this way, new aspects are discovered, perhaps unknown in the working relationship, and it becomes possible to make an appropriate decision for both sides.

Within the process of technology transfer, financing plays a fundamental role in bringing scientific developments from the laboratory to the market. When choosing to undertake and create a new science-based company, it will be necessary to be prepared to travel the demanding path of seeking financing. Depending on the development phase of the venture, entrepreneurs will need to knock on different doors. In turn, each phase will have different strategies and arguments for raising the much-desired capital.

In the initial phase, or the creation of the science-based venture, financing comes mainly from government subsidies; often the entrepreneurs themselves assume some of the costs, although, in some cases, there may be third parties willing to invest. The argument to convince the third party, at this stage, is based almost exclusively on a hypothesis.

In the following stage, the weight of the hypotheses remains very significant. However, some validated metrics begin to appear, which helps convince the third party to invest. In this sense, the friends, family, fools, and founders (4F Model), or as it was renamed here, friends, family, founders, and smart (3F + S Model), are those who are more available to invest.

In the next phase, validated metrics begin to gain more prominence in convincing someone to invest, although the proposed hypothesis remains more relevant. Government subsidies, angel investors, and family offices begin to appear at this stage.

In the following phase, validated metrics gain more value than the hypothesis when raising investment capital as a history of metrics begins to be established. Those willing to invest in this stage continue to be the government through subsidies, as well as crowdfunding and venture capital.

Finally, in more advanced stages, validated metrics are the main argument that will allow raising investment capital, and the hypothesis will be marginal. Investments are most likely to come from corporate venture capital and, at an advanced stage, private equity.

The "Death Valley"

The "death valley" is an expression commonly used in the world of innovation to describe a stage that every venture must overcome, related to the lack of money or negative cash flow during the early stages of development.

There are studies (Gompers & Lerner, 2001) that state that 90% of ventures do not manage to raise investment in the first three years of existence. In this phase, they usually go through the death valley, where ventures are in constant danger as they must make permanent efforts to survive and not run out of liquidity. At this stage, third-party financing takes on special relevance, which means that ventures developed in environments more conducive to taking risks have a higher probability of success. In the Latin American case, companies that overcome the death valley do not represent a percentage higher than 8% (Musso, 2012).

However, according to what was proposed by Karen Gullbrandsen from Iowa State University, in technology transfer, the "valley-of-death" metaphor is often used to describe the gap between academic innovations and their arrival in the market. The more traditional definitions of technology transfer indicate that the transit of intellectual property from university research laboratories to private companies or public institutions that commercially develop the technology is rather smooth and with low risks. The death valley presents the opposite view. In fact, this rather gloomy metaphor implies that academic research is somehow isolated from the outside world, and often, there is the realization that solutions are being proposed to problems that do not exist (Gulbrandsen, 2009).

Although the death valley suggests that technology transfer is about moving ideas from one entity to the next, technology transfer is about linking people, relationships, and collaboration between individuals and groups (industry, government, and academia) with varied interests (Gulbrandsen, 2009).

References

Ahn, M. J., & Meeks, M. (2008). Building a conducive environment for life science-based entrepreneurship and industry clusters. *Journal of Commercial Biotechnology, 14*(1), 20–30. https://doi.org/10.1057/palgrave.jcb.3050076

Carbonell, S. D. T. (2012). Financing sources for innovators in Argentina: Venture capital, angel investors, institutional investors. *Revista de La Bolsa de Comercio de Rosario*, 50–57.

Gompers, P., Kaplan, S. N., & Mukharlyamov, V. (2015). *What do private equity firms say they do?* (pp. 1–23).
Gompers, P., & Lerner, J. (2001). The venture capital revolution. *Journal of Economic Perspectives, 15*(2), 145–168. https://doi.org/10.1257/jep.15.2.145
Gulbrandsen, K. E. (2009). *Bridging the valley of death: The rhetoric of technology transfer*. Iowa State University, Digital Repository. https://doi.org/10.31274/etd-180810-527
Musso, R. (2012). *The valley of death.*
Sethi, A. (2016). from science to startup. In *From science to startup: The inside track of technology entrepreneurship*. Springer International Publishing. https://doi.org/10.1007/978-3-319-30424-3
Steve, V. (2014). *Vicente Esteve | Finance for startups*. https://vicentesteve.com/
Steve, V. (2017). *FFF financing for startups | Vicente Esteve*. https://vicentesteve.com/financiacion-fff-para-startups/

7 Technology Transfer Management

Abstract

This chapter addresses technology transfer management and technology portfolio management. It involves managing a set of research outcomes, establishing entry criteria, and reducing variability. This facilitates reporting, analysis, and better management of technologies. For active portfolio management, there are different methodologies, such as the quadrant methodology, which evaluates the technological and commercial potential of each technology, and the level methodology, which classifies technologies according to their level of impact. In addition, the tool that measures the maturity level of technologies (TRL) is detailed. Finally, it is argued that technological marketing is essential in the technology transfer process and focuses on solving problems and meeting the needs of the industry. Conducting proper market research and adapting the message to the target audience are key aspects of technological marketing.

Keywords

Technology transfer management • Technology commercialization • Science-based ventures • Spin-off • Technology transfer office • Market validation • Technology portfolio management • Technological marketing

General Aspects of Technology Transfer Management

"Technology transfer management" corresponds to the actions carried out in the technology commercialization process. It is structured in four key stages when bringing a research outcome to market. The first stage is research, the second is commercial validation, the third is intellectual property management, and the fourth is technology commercialization.

A. Ossa, *From Research to Market*, https://doi.org/10.1007/978-3-031-71340-8_7

The first stage consists of the research itself. However, the technology transfer process begins with the research outcome, which must be officially communicated to the office that manages this activity within the institution, often through a document called "invention disclosure." At this stage, offices also manage funding for research, which comes mainly from the government in the early stages.

Once the research outcome is identified, a feasibility analysis or technical and commercial validation must be carried out, which, if appropriate, structure the protection and transfer strategy.

The third stage corresponds to intellectual property protection. The protection strategy determines the most appropriate mechanism to safeguard the research outcome in accordance with the commercialization strategy.

The fourth stage is the execution of the commercial strategy, which has been previously established, at least preliminarily, and which allows the research outcomes to reach the market and society, either through a license agreement with a third party or by creating a technology-based venture, also called a "spin-off."

Technology Transfer Process

The technology transfer process is quite standard, and with it, the functions of a technology transfer office or the units responsible for managing technology commercialization do not vary much from one institution to another. However, there are some nuances of integration backward or forward, depending on the reality of each institution and country.

The following is a model based on "The 8 steps of technology commercialization" from the University of Texas (*Process of Technology Commercialization*, 2016), called "Technology Transfer Cycle" for the reality of Latin American countries or those with similar characteristics to this region.

The first step of the cycle is the management of resources for research and development, called "R&D Management." Resources can come from the government or the private sector. In those economies that do not have a sophisticated economic and innovation system, resources are mainly government-funded, versus those more developed economies where around 70% come from the private sector. In the case of government resources, funding sources must be identified, and the necessary steps must be taken to apply for the funds, which are mostly subsidies. Private financing, usually known as R&D contracts or "contract research," is negotiated directly with the financier(s), and since it is a direct deal, it is adjusted according to individual negotiations. This is a recurring process until the necessary funding is found to carry out the research, which can be recursive until a research result is attained.

The second step is called "Search for R&D results." Specifically, it refers to scouting research results within the institution because researchers often obtain results with high transfer potential, but they do not know it. This scouting management must be carried out constantly and systematically. Once a research result is found, the researcher must make the invention declaration, completing the document indicating the title of the invention, name of the principal investigator, impact area, projects that originated the invention, whether or not third-party materials were used, description of the invention, problems it solves or potential applications of the technology, among other points. If the researcher does not yet have all the information or has not yet completed the development, they can submit the declaration form with the information they have and this is moving on to the next step. During the process, the members of the transfer office must win over researchers who are not interested in technology commercialization and who often do not want to embark on this challenge simply due to a lack of knowledge.

The third step of the cycle is called "early commercial validation." The aim is to validate with the corresponding key actors that there is indeed a problem or need, that this problem is relevant, and that the proposed solution is perceived as adequate. It is well-known that many of the technologies developed by universities or research centers are carried out from scientific curiosity. Among other things, it is important to validate early on whether there is a real market need for what has been developed, under what conditions it could be commercialized, and the critical factors necessary for this to happen. It is essential to carry out these early validations with key industry actors, understanding the existing problem or need in depth. It is highly recommended that those who carry out these validations have technical knowledge of the solutions and can understand the commercial language. It is a fundamental requirement that they understand in depth the technological development they wish to validate. This is one of the most crucial steps in the transfer process and is key for what comes later. It may happen that the commercial validation is not favorable, making it necessary to return to the research process or, alternatively, iterate and modify the proposal to be validated until it is of interest to the industry.

The fourth step is "intellectual protection." The most relevant aspect at this point is to develop the best intellectual protection strategy (as opposed to the specific management of protection already in place), because the greatest value of protection lies in finding the mechanisms that bring greater robustness to a technology. In some cases, offices must act as intermediaries between the research team and external law firms that manage the protection of research results, maintaining internal management for some things; in other cases, offices carry out all management internally; and in still other cases, it depends on the type of protection. For example, if it is a patent application, it is likely to be outsourced to a law firm. In contrast, if it is a copyright registration or industrial secret protection, it is possible that the management is carried out internally. If more research or results are needed to support the hypothesis, it may not be possible to achieve an adequate protection strategy. In that event, it is necessary to return to the first step.

The fifth step is "valuation and prospecting." At this stage, relevant actors, which allows a deep iteration, are carried out to guide the team in developing a value proposition and consolidating the model and value proposition for a potential licensee. After this, the search for partners begins, and it is necessary to have a clear value proposition. Validated technology, intellectual protection initiated, and all actors aligned with the process allow the technology to be valued. It is necessary to proceed as best as possible with the resources and information available up to this point. This is probably one of the most complex stages of the entire process, which is why the role of the technology transfer office is crucial. Although it is true that most potential partners start as the researcher's initial contacts, the transfer office is dedicated to better materializing the agreement, mediating in the negotiation, and providing greater guarantees to the third party that the partnership is a good alternative for both parties. Many technologies do not find partners, so it is necessary to return to the research process to strengthen the result enhance other applications for the same technology or, alternatively, simply abandon the idea and embark on a new challenge. The intellectual property strategy could also need to be strengthened.

The sixth step is "negotiation and signing" of the license agreement. At this stage, the terms must be negotiated with the interested parties and materialized in a contract, paying attention to both form and substance aspects. In this step, negotiation skills, team building (to-face negotiation), and guidance from team members as well as external parties (if necessary) are key. Being very proactive and responding to industry timelines is recommended. This stage involves legal matters, and lawyers play an important role. It is essential that they are aligned with the entire process and understand the context to avoid basic mistakes, e.g., trying to negotiate already agreed-upon points or questioning established agreements. It is possible that no agreement is reached, in which case it would be necessary to return to the previous step.

Step 7 is called "follow-up." The transfer process does not end with the signing of the license; rather, it is the beginning of a relationship with a new partner. Once the license is signed, the agreed-upon plan between researchers and partners must be followed, including potential revenues. Often, new R&D management opportunities may arise from this relationship, meaning that it could return to the first step or even have to renegotiate the licensing terms due to changes in the environment.

The main reason for licensing a technology is that the solution can reach the market and solve a real problem in the industry and society. Generating revenue for both the research team and the institution under which the research was developed can also be a legitimate goal. There may also be good political and strategic reasons for signing a license agreement.

The research team should generally remain involved in the process once the license agreement is signed. Therefore, it is important that the inventor, the university, and the licensee have a good working relationship both during negotiations and after signing.

Science-Based Entrepreneurship Creation Process

The process of creating science-based ventures requires specialized support, and only some technology transfer offices have developed this capability. Science-based entrepreneurship is a transfer vehicle, but it is not the most suitable vehicle in all cases. To establish the needed specialized support, dedicated work from researchers and the technology transfer office team is required. Since private capital is necessary for development, the challenges are even greater, requiring experience, networks, and credibility—elements that can be provided by the technology transfer office. These offices should also promote the formation of teams for venture development, with mixed profiles between researchers and entrepreneurs, in order to have the different capabilities to carry out an endeavor of this magnitude.

Different universities have support for the creation of "spin-offs," either through their technology transfer offices, business incubators, or even accelerators. The services they provide range from business plan development to industry contacts. The following are some examples. Ghent University, Belgium has two types of support: one for traditional ventures and another for science and technology-based ventures. Their services range from communicating the various websites that can provide information to support creating a spin-off, to helping find funding. Their main feature is that they have a contact person for those who want to start a business, so they can be contacted and start the process. Ku Leuven University in the same country provides intellectual property management services, legal support, advice on business plan development, and support in validating the business model and market, as well as helping to find a competent team for venture development. Carnegie Mellon University in the United States has a specialized area for creating spin-offs called the Center for Technology Transfer and Enterprise Creation. In addition to providing the services mentioned by other universities, the Center helps find physical spaces for incubating the spin-off. The Instituto Tecnológico Superior de Monterrey leads Mexico's example, which has transfer policies and programs that include a competition supported by government resources, in which academics who want to start a business have access to both methodological tools and funding to develop their spin-off. In addition, they are given the opportunity to dedicate 25–50% of their research during a year to transfer and entrepreneurship activities. Furthermore, this university has networks of technology parks that further enhance its transfer work.

At the Pontificia Universidad Católica de Chile, an academic entrepreneurship program has been created, which aims to systematically support the creation of spin-offs. The purpose is to contribute to society through the creation of science-based ventures that solve problems of national and global relevance, as well as to articulate and develop the university's capacities to support the formation, acceleration, and capital raising of sustainable and high-impact research-based ventures for society. This is pursued by identifying the different actors of the university involved in the development and formation of spin-offs, by providing support in the form of advice, connections, and value-added services to those who wish to form science-based companies, and by promoting an entrepreneurial culture in academic and research activities.

The support prepares for the formation of the spin-off by strengthening the protection strategy, the business model, the value proposition, and the business plan. Then, the terms of the technology licensing are agreed upon, and the valuation and preparation of the "pitch" are carried out to start negotiations with potential investors. On the other hand, connections are promoted with experts who can guide the technology, with companies, and with investment funds. Support for strengthening entrepreneurial teams is essential, so the program is oriented towards training researchers and conducting searchs for someone who could lead the management and commercial aspects of the venture, called the "champion."

The process is divided into four stages (Sullivan, 1995). In the first stage, a formal declaration is made by the research team indicating that they want to form a science-based venture. This is crucial, as the process must be voluntary, and it is best to work only with those who are truly interested in embarking on this challenge. A working team is assigned to accompany the process, which begins with a diagnosis and the development of a preliminary work plan. This stage should last between six and twelve months, and at the end of it, a presentation is made to an external panel, which then suggests whether or not to move on to the next stage.

In the second stage, work is done alongside a strategic committee that supports monitoring the work plan's execution and provides continuous feedback. In this phase, the entrepreneurial team is intensively strengthened. At the end of this stage, a presentation is made to the institution's spin-off board, and the decision is made to move on to the next stage, which includes a seed capital contribution of $20,000. This stage should last six months.

In the third stage, the business model is further developed, the company's valuation is carried out, and continuous monitoring is conducted by the strategic committee, before finally being presented to the spin-off board, which approves the formation of the spin-off, and injects $80,000 which, together with the initial seed capital, are become convertible notes when investment is raised and the venture is legally formed. This stage should last from six to twelve months.

In stage four, the focus is on raising private capital: presentations are made to potential investors, the company is legally established, and a dissemination plan is developed. This stage lasts one year, and by the end of this stage, the venture should be ready to enter the market (Sullivan, 1995).

Portfolio Management

Technology portfolio management is very relevant within the technology transfer process; it refers to the administration of a broad group of research outcomes and considers the overall picture of all technologies in order to prioritize and coordinate the work of each one, thus optimizing the task.

The technology portfolio should consist of all applied science projects being executed and expected to generate an appropriable and potentially transferable result. In some cases, it may be efficient to divide the portfolio into subgroups (using the scope of the technology's impact as the grouping criterion), since the protection and transfer strategies for these results—as well as the language necessary to reach research teams and other stakeholders—vary considerably depending on the scope.

To begin the formation of the portfolio, it is important to determine the criteria for a technology's entry into it and reduce the variability of criteria. This allows for standardization not only of data entry but also of the work that begins from there, facilitating better reports, variable crossings, and overall better management of technologies.

Portfolio management is carried out in a structured manner. There are also opportunities to review prioritization, depending on the potential of the result in response to certain criteria. To define a good strategy around a research result and to prioritize efforts, it is essential to carry out a strategic analysis of technical-commercial feasibility focused on six criteria: the state of technology development, the market analysis carried out, the financing required to bring the technology to market, the regulatory aspects with which the technology must comply, the intellectual property strategy, and the commitment and competencies of the team.

Although all criteria are vital and interrelated, the criterion referring to the team's commitment is perhaps the most strategically important. This is because, first, technology without a team to accompany and push the transfer effectively never reaches the transfer stage; and second, because generally, technologies must be jointly adapted or worked on in order to achieve a minimum viable product. This requires, without exception, a team that communicates with other stakeholders, is open to diverse opinions and interests, and is committed to the importance of the transfer.

It is important to classify the portfolio. While including the total number of technologies, it is recommended that they be divided by active and inactive status. The inactive portfolio is the one that does not require management, while the active portfolio is the group of technologies that require transfer management in the short, medium, or long term.

Technologies in the inactive portfolio do not have or require management for a variety of reasons: the research team either lost interest in the development or did not achieve protectable or transferable results; the technology received a negative evaluation from the industry, the proposal needs to be redirected, but the research team is not interested in doing so; or, there is no interest from a third party in commercially exploiting the technology. In the future, these initiatives can become active again if the research team expresses formal interest in continuing the work, so it is essential that inactive technologies are not removed from the portfolio.

There are a variety of management methodologies for the active portfolio, which depend on the portfolio size, work structure, and, above all, the management strategy.

Quadrant Methodology

The quadrant methodology, developed at the Pontifical Catholic University of Chile and based on different portfolio management processes of entities such as the University of Campinas, SRI International, and the University of California, consists of prioritizing technologies. It is used to classify and identify those needing to advance in research development, search for new markets, and research results ready to transfer. When evaluating the portfolio, technologies are positioned on a Cartesian plane, that is, two perpendicular numerical lines, one horizontal and the other vertical, which intersect at a point called "origin" or "zero point"; for this reason, it is called the "quadrant methodology." The technological potential is evaluated on the horizontal axis, and the commercial potential on the vertical axis.

This tool was developed to prioritize research results and focus transfer efforts on the highest-potential R&D results.

First, an evaluation of each technology is carried out, rating and assigning a score to each R&D result according to previously established criteria, which in turn are made up of subcriteria. All scores range from 1 to 5, and each criterion and sub-criterion is weighted to construct the final score. Four criteria are evaluated: the invention, intellectual property, the market, and the research team. The sub-criteria are more specific aspects that build the score of each criterion.

First criterion: the invention.

The invention criterion is evaluated on three sub-criteria: the degree of innovation, the degree of development, and the time to reach a precompetitive prototype.

The degree of innovation has a weighting of 20% and is rated as follows: (1) if it has no inventive level; (2) if it exhibits a small improvement of something already existing; (3) if the development is a significant improvement or allows a second use; (4) if it is a new process or product or a new use very different from the original use; and (5) if it is a disruptive invention in the state of development.

The degree of development has a weighting of 60%. A rating of (1) is given if the development is at the conceptual stage; (2) if the innovation is at the stage of validation of scientific hypothesis; (3) if it is at the stage of prototype development at the laboratory level; (4) if the innovation is at the stage of product development or pre-commercial; and, (5) if the technology is at the packaging stage.

Finally, the time to reach a precompetitive prototype, is weighted at 20%. It is rated as (1) if the probable wait period is three or more years; (2) if the period is two to three years; (3) if it is two years; (4) when the probable period is between one and two years; and (5) if the development will reach the prototype in less than a year.

Second criterion: intellectual property.

In criterion 2, intellectual property aspects are evaluated. The first sub-criterion is the ownership of intellectual property, which is weighted with 20%. It is rated (1) if there is no absolute clarity of the ownership of the intellectual property; (2) if there is clarity of the ownership, but it is not formalized; (3) if ownership is shared among three or more owners; (4) if ownership is shared with a third party; and (5) if the institution has 100% of the ownership.

The second sub-criterion is the origin of the development, and it is weighted with 20%. It is rated (1) if the development used material or information from third parties without clear conditions of use; (2) if the development used material or information from third parties with authorization but with certain restrictions; (3) if the development used material or information from third parties with authorization that allows commercial use, but under certain conditions; (4) if the development used material or information from third parties with authorization and without conditions; and (5) if the development used material or information from third parties for free use or in the public domain.

The third and final sub-criterion, the degree of appropriability, is weighted by 60%. It is rated (1) if it is very easy to infringe intellectual property rights or to copy the obtained solution, as no technical competence or third party's right is required for its use; (2) if it is easy to infringe intellectual property rights or copy the obtained solution, as basic technical competencies are required; (3) if by using published information or obtained from research results, it is possible to develop a similar or second-use solution; (4) if intellectual property rights could be infringed using published information but adding technical competencies; and (5) if it is very difficult to infringe intellectual property rights or to copy the solution, as in addition to information, deep knowledge on the subject is required.

Third criterion: the market.

Criterion 3 is the market, and it has three sub-criteria. The first sub-criterion is the definition of the solution, which has a weighting of 10% and is rated (1) if there is no definition of the product; (2) if there is a vague definition of the product; (3) if a preliminary definition of the product is displayed, although it is not clear how it will be marketed; (4) if there is a preliminary definition of the product and a preliminary business model; and (5) if a finished definition of the product or solution, a business model, and a defined packaging mechanism are displayed.

The second sub-criterion is the market potential for the solution, which is weighted at 40%. It is rated (1) if the product has no market demand; (2) if the product exhibits a limited potential demand or could meet needs in a niche market; (3) if the product exhibits medium potential, could meet needs, and compete in a limited segment; (4) if the product exhibits high potential demand and could meet needs in a massive segment; and (5) if the product has high potential demand and could compete in a global market.

The third sub-criterion is the potential product value. It is weighted with 10% and is rated (1) if it addresses a low-priority issue and its solution impacts a small number of users; (2) if the technology addresses a medium-priority issue and its solution, impacts a small number of users; (3) if the development addresses a medium-priority issue and its solution, impacts a significant number of users; (4) if it addresses a high-priority issue and its solution impacts a significant number of users; and (5) if the technology addresses a high-priority issue and its solution impacts a large number of users, global market.

The fourth sub-criterion is the degree of substitution of the product by emerging solutions. It is weighted with 40% and is rated (1) if there is a high risk of substitution or alternative solutions already exist that have clear competitive advantages; (2) if there is a normal risk of substitution or alternative solutions exist that exhibit some important differentiating factors; (3) if there is a normal risk of substitution or alternative or emerging solutions exist, although it was not possible to determine if they have a key differentiating factor; (4) if the risk is normal-low of substitution, or if emerging solutions exist but they clearly do not have competitive advantages; and (5) if the risk of substitution is marginal and it was not possible to determine if there are emerging or alternative solutions with clear advantages.

Fourth criterion: the research team.

Criterion 4 is the research team. Sub-criterion 1 is the team's experience in transfer processes. It is weighted with 10% and is rated (1) if it has no experience; (2) if it has no experience but understands the process; (3) if at least one team member has indirect practical experience; (4) if at least one team member has direct practical experience; and (5) if more than one member has direct practical experience.

The second sub-criterion is the team's reward expectation. It is weighted with 10% and is rated (1) if reward expectations are unattainable; (2) if reward expectations are hardly attainable; (3) if reward expectations are coherent; (4) if reward expectations are possibly attainable; and (5) if reward expectations are highly attainable.

The third sub-criterion is the public–private networks of the research team. It is weighted at 15% and is scored (1) if they do not have networks; (2) if they have at least one contact but are not key players; (3) if they have more than one contact, but are not key players; (4) if they have more than one contact, and these are key players; and (5) if they have a network of contacts.

The fourth sub-criterion is the interest of the research team in continuing with the research line. It is weighted at 15% in the evaluation process. A score of (1) is assigned if the research team explicitly states their intention not to continue with the research line. A score of (2) is assigned if the research team expresses a belief that they will not pursue the research line. If the research team has not indicated regarding their intent to continue with the research line, a score of (3) is given. A score of (4) is assigned if the research team believes they will continue with the research line. Lastly, a score of (5) is given if the research team explicitly states their commitment to continuing the research line.

The fifth sub-criterion is the team's publications regarding the technological solution. It is weighted at 10% and is scored (1) if they have not yet published; (2) if they have published fewer than five articles in low-impact journals; (3) if they have published at least five articles, but in low-impact journals; (4) if they have published at least five articles in high-impact journals; and (5) if they have published more than five articles in high-impact journals.

The sixth sub-criterion is the team's availability to participate in the transfer process and is weighted at 40%. It is scored (1) if they do not participate in these tasks; (2) if they do not believe they have time to accompany these tasks; (3) if they have not expressed their willingness; (4) if they believe they have time to accompany these tasks; and (5) if they have stated they are 100% available to participate in the process.

The last step in the quadrant methodology is to construct a two-axis graph where the horizontal axis is the technological potential and the vertical axis is the commercial potential. The scale will be from 0 to 5 with intervals of 0.5, and the axes will intersect at point 2.5. The technological potential will be calculated as an average between the values of the invention and intellectual property criteria, and the commercial potential will be calculated as an average of the values obtained in the market and research team criteria.

R&D results located in the upper right quadrant are those with the highest potential and require the greatest efforts for their transfer. Technologies located in the upper left quadrant have a high market potential but either lack R&D, the technology is weak or still has many technological questions or risks, or the intellectual property is weak. This indicates the importance of focusing on developing the technology further in these cases and clearing up those questions.

R&D results located in the lower right quadrant, on the other hand, are robust technologies but have no market. This indicates that the focus should be on finding other applications for the same technology, such as expanding its uses or applications.

Those technologies found in the lower left quadrant have low commercial potential and weak technological development, so allocating efforts in managing their transfer is not recommended.

Levels Methodology

Developed at the Pontifical Catholic University of Chile, this methodology is based on the classification of technologies according to their relevance. Two criteria are used: the level of impact, and the potential for commercialization. That is, the levels methodology is not based on the entry conditions of the technologies, but rather on the exit conditions.

The potential for commercialization is measured by analyzing the problem and the solution, as well as the structuring of the value proposition. However, validation with key industry players is essential to determine this potential.

The other criterion is the level of impact. The impact can be economic, social, environmental, or cultural. However, the key is how the institution defines it. Here are some examples of how to define this criterion: it significantly and unquestionably increases the value of a good or service in the market; the technology causes a positive disruption in the behavior of one or more industries. It generates high income for the institution; it solves a problem or generates a change in sustainability, environmentally or otherwise; it significantly and unquestionably improves the quality of life of a significant number of people or of a minority group. It could also be something that becomes a "gold standard," which refers to something that is very good and is used to measure how good other similar things are.

This methodology divides the active portfolio into four levels, and level 4 is further divided into four sublevels. Level 1 is for all those technologies with low transfer potential and, usually, low potential impact. The strategy for technologies at this level is more about monitoring their progress over time, being attentive to opportunities or changes in environmental conditions that may affect their potential for both impact and transfer, such as regulatory changes or potential new applications that arise for a technology.

Level 2 includes all those technologies with the possibility of being transferred to the market but with low or moderate impact. The strategy in these cases is to generate proactive transfer actions, but with limited efforts if the capacities and resources are also limited. It is very likely that this level concentrates on a high level of technologies, so a strategy of transferring a large number of technologies is developed rather than focusing on the impact that each one can generate.

Level 3 includes all those technologies that have a high probability of being commercialized and can generate high impact at the same time. Level 3 strategy is of great effort and focus because, although it is true that the volume is likely to be much lower than in the previous level, the level of impact that each individual technology can generate may be much greater than several transferred technologies from level 2.

Level 4 is comprised of all those technologies that have already been transferred to the market. Unlike the previous levels, this one is divided into four sublevels: sublevel 4.1 corresponds to those technologies that have already been non-exclusively licensed (i.e., they can be licensed again), but the impact is low and with low possibility of licensing to another entity; sublevel 4.2 includes those low or medium impact technologies that are non-exclusively licensed, and have the possibility of being licensed to another entity; sublevel 4.3 corresponds to high-impact technologies that are non-exclusively licensed and with the possibility of licensing to another entity; sublevel 4.4 includes those licensed technologies, with no possibility of licensing to another entity, as exclusive contracts have been signed.

In the methodology of levels, these some factors are not used for the classification process but are relevant in the commercialization strategy: the state of development; the regulatory environment; intellectual property aspects; and the research team.

The state of development is not a determining factor in portfolio prioritization; that is, there may be technologies that are in the initial stages of development or in more advanced stages, yet are classified at the same level. The strategy to implement concerning the state of development is to carry out the relevant actions for the progress of the technologies.

Likewise, the regulatory environment does not influence the classification of technology at one level or another, as it seeks to work with all technologies so that they can meet market demands in regulatory terms.

In intellectual property aspects, relevant factors for the progress of the technology must be considered, including freedom to operate and intellectual property strategy. In this factor, the protection mechanism does not matter; rather, developing a robust protection strategy is important.

Of primary importance regarding the research team is to adequately diagnosis the initial situation, then evaluate the best mechanisms to reinforce those aspects that allows the formation of a group proficiently capable of carrying out the technology transfer.

For this methodology, the classification process by levels is carried out with what is called "expert criterion." A group of experts, either internal or external to the institution, are consulted about which classification level they believe is most suitable for a certain technology, which allows for validating the initial proposal generated by the person in charge of the research result. This mechanism is normally used to solve complex problems, giving independence and peace of mind to the participating experts. In turn, this allows for different perspectives to be combined in order to reduce error and bias.

The use of guidelines can make this classification more objective. In order to build the rubric, it is very important to start by knowing the actual state of development—not to use it as a classification method, but to estimate whether the information available is more of an estimate due to the preliminary level of development, or if it is already based on a proven hypothesis, as in the case of more advanced developments. It is also necessary to analyze the advantages of the technology compared to existing developments and the approach to the problem or opportunity that the technology solves. In regulatory terms, it is necessary to know the current regulations, to know the validations carried out, as well as to understand the real market interest and the possible applications that the technology could have, in addition to understanding the research team's capacities, both technical and commercial. Likewise, it is important to understand, in terms of the technical implementation of the transfer, what is required for the technology to be transferred to the counterpart and for the counterpart to be able to use it, that is, if it is only necessary to deliver the patent or if it is necessary to transmit *know-how*, or if technical assistance is required for its implementation. Regarding intellectual property, it is necessary to know if there is freedom to operate or if there are third parties with rights. All of the above allows for the estimation of whether the technology has commercialization potential and the evaluation of the potential level of impact.

It is important, for the methodology to work, not to establish a minimum or maximum number of technologies per level. Simply select all those that can meet the requirements. This becomes relevant when making an adequate classification, strictly using commercial criteria and potential impact without other factors influencing, such as quotas per level.

Strategically, the technologies to which greater commercialization efforts should be dedicated are those at levels 3 and 4.3; these have the highest probabilities of commercialization and high impact. Additionally, since the technologies classified at level 4.3 have already been transferred, the possibility of transferring to another third party, theoretically, is greater. Secondly, efforts should be made to transfer technologies classified at levels 2 and 4.2 since, with transfer possibilities, the level of impact is low or medium. Finally, commercial efforts in technologies classified at levels 4.1 and level 1 should be much lower.

Technologies can change levels depending on their evolution, which could cause them to rise or fall. It is advisable to build a twelve or twenty-four month roadmap for each technology, and to be attentive to their evolution. Periodic reviews every three or four months are suggested.

It may be that a technology initially has a high potential for commercialization, but the commercial validation with the industry turns out to be negative, and there are no capacities to redirect the technology. Or it may be that some regulatory change facilitates the transfer of a technology, which, having had high potential, had low transfer possibilities.

Both methodologies proposed above can be combined with the "funnel method." This method is the most well-known. It is so-named because, on its wider side, many technologies enter; however, few manage to exit through the narrower end and complete the journey for technological development to become an innovation. In the process, certain requirements must be met to move on to the next stage, called "staged gate," where it is decided whether the technology advances to the next stage or leaves the process. This decision must be based on previously established criteria; however, in this method, the most relevant factor is the level of technology development. Therefore, it can be complementary to quadrant methodology and level methodology as portfolio management tools.

At the development stage, technology is in the research phase; it must advance from the conceptual idea to validation at the proof of concept or initial prototype level in the laboratory, which allows for verifying the scientific hypothesis and, at the same time, reduces the risk involved in the research. The protection of research results by intellectual property rights mainly occurs at this stage. Additionally, at this stage where the connection with industry begins, to either validate the problem and solution, joint development or certify the research results obtained.

The packaging and scaling stage begins with a result at the proof of concept or functional prototype level, generally protected by intellectual property rights, and culminates with a validated and packaged prototype that has the minimum features to be transferred. Depending on the technology development plan, this could be the stage to apply for competitive funds, or to seek public or private financing to refine the prototype, package, and scale the technology in a test environment. It is expected that at this phase the technology begin to advance towards a transferable good or service. At this stage, it is also beneficial that the technologies have an external partner interested in the commercial exploitation or massification of the research results.

The transfer stage begins with protected and validated results in an environment relevant to the industry and, ideally, with part or all of the actors in the production chain. During this stage, a risk analysis is carried out, and a transfer plan is defined, which may involve development stages, clinical studies, or other relevant activities. At this stage, potential partners are prospected, and the transfer conditions are negotiated until they are finalized.

It is essential to understand that these models must be adapted, depending on the type of technology. For example, biomedical developments are generally transferred at earlier stages and not after a clinical phase 1 (where there is still a long way to go to reach the market), versus software that usually be transferred when it is already operational.

Technology Maturity Levels

The "technology readiness levels" (TRL) (Ibáñez, 2014) is a methodology developed by the National Aeronautics and Space Administration, better known as NASA. It was developed during the space race in the 1970s, seeking to establish a standard to estimate the maturity of a technology, with the aim of facilitating its acquisition.

For some decades, it has become popular in technological development worldwide. For example, at the beginning of the 2000s, the United States Department of Defense began to use it; at the end of that decade, the European Space Agency (ESA) did the same, and in 2010 the European Commission advised its use in research projects.

Although the initial purpose of the TRL scale was linked to space use and was usually used to facilitate decision-making related to technology development and transition, it is currently used as a tool to measure the evolution of research and development. Among the main advantages is that a standard and known mechanism is established to evaluate the state of development of a technology. However, among the disadvantages is that the maturity of the technology does not necessarily guarantee its usefulness, as multiple environmental factors influence this last factor (Ibáñez, 2014).

The scale originally used by NASA included only seven levels:

Level 1—Basic principles observed and reported.
Level 2—Validated potential application.
Level 3—Proof of concept demonstrated analytically and/or experimentally.
Level 4—Component validated in the laboratory.
Level 5—Component validated in real or simulated space.
Level 6—System adequacy validated in a simulated environment.
Level 7—System adequacy validated in space.

In the 1990s, the other two levels were incorporated, and the definitions initially proposed by the aerospace organization were adapted:

TRL 1—Basic principles observed and documented.
TRL 2—Technology concept and/or application formulated.
TRL 3—Critical function proof of concept demonstrated analytically and experimentally and/or characteristic.
TRL 4—Component and/or test board validation in laboratory environments.
TRL 5—Component and/or test board validation in a relevant environment.
TRL 6—System/subsystem model or prototype demonstration in a relevant environment (ground or space).
TRL 7—System prototype demonstration in a space environment.
TRL 8—Completed real system and "flight qualified" through test and demonstration (ground or space).
TRL 9—Real system "flight tested" through successful mission operations.

On the other hand, the scale used by the European Union is as follows:

TRL 1—Basic principles observed.
TRL 2—Technology concept formulated.
TRL 3—Experimental or concept test.
TRL 4—Technology validated in the laboratory.
TRL 5—Technology validated in a relevant environment.
TRL 6—Technology demonstrated in a relevant environment (industrially relevant environment in the case of key enabling technologies).
TRL 7—System prototype demonstration in an operational environment.
TRL 8—Complete and qualified system.
TRL 9—Real system tested in the operational environment (competitive manufacturing in the case of key enabling technologies, or in space).

In the case of research and development, some clarifications have been made, even incorporating the TRL 0 level:

TRL 0—Idea. Unproven concept, no tests have been performed.
TRL 1—Basic research. Postulated and observed principles, but no experimental evidence is available.
TRL 2—Technology formulation. Concepts and hypotheses have been formulated.
TRL 3—Applied research. First laboratory tests completed and proof of concept performed.
TRL 4—Small-scale prototype performed in the laboratory.
TRL 5—Larger-scale prototype tested in the field.
TRL 6—Prototype system tested in the field with performance close to expected.
TRL 7—Pre-commercial demonstrative system operating in an operational environment.
TRL 8—First commercial version. Manufacturing and design issues resolved.
TRL 9—Full commercial application. Technology available to consumers.

From TRL 0 to TRL 4, activities are carried out in the laboratory environment and are commonly known as the research stage. On the other hand, in TRL 5 and TRL 6, activities are carried out in a simulation environment, also called a "relevant environment," as it refers to those conditions that approximate or simulate the environment where the technology ultimately be used. This stage is also known as the development phase. Finally, TRL 7, TRL 8, and TRL 9 correspond to all activities carried out in a real environment. Since the technology is already at the market level, it is known as the innovation phase (Ibáñez, 2014).

Technological Marketing

In the context of technology transfer, "technological marketing" is the process of making companies interested in the technologies developed by the institution and, just like traditional marketing, it requires knowing and analyzing the market, as well as the interests of customers.

Digital marketing is also known as *technological marketing*. In that case, it means using technological tools to market a product, not to market of technological developments. They are absolutely different concepts.

This role is often undervalued within the technology transfer process and, in turn, is confused with the task of only promoting technologies. Just as in traditional marketing, advertising is only one element of the process since "**marketing** focuses on creating a market for the product and building a reputation, while **advertising** focuses on seeking public attention" (*Do you know the difference between Marketing and Advertising?—More Than Business.* 2018).

A key role of technological marketing is the identification and management of key actors according to the different stages of the technology transfer process. In the R&D stage, it is relevant to maintain relationships with universities, laboratories, and research centers from both the public and private sectors. In the development stage, private companies and government agencies is essential. In the final phase, primary consumers, that is, end-users, and secondary consumers, those individuals who buy and recommend, or also service providers, is key.

Finally, there are permanently relevant actors throughout the technology transfer process, such as government agencies that can contribute with subsidies or even be clients, private companies, brokers, venture capitalists and angel investors (Bauer & Flagg, 2010).

The consulting firm 4 innovation (Canetti, 2012) provides a comprehensive analysis of the technological marketing process; it states that selling is the art of helping someone get something they need, or find a solution to their problem. When it comes to research and technological development, it is very relevant to address the interests and know the needs of the industry to guide technological developments.

In this sense, selling is solving problems such as expanding product lines, reducing manufacturing costs, or increasing profit margins and market share quotas. To generate a sale, it is necessary to speak the language of the industry and focus on the benefits, but not necessarily on its features, of the technology. Features are specific aspects that can be proven, but they are not attractive for selling or convincing someone to buy something. On the other hand, benefits are aspects related to the needs of customers, with a clear value for them, and are the reason why they make the decision to buy; benefits are what will convince customers to buy (Canetti, 2012).

In the case of technology transfer, the purchase is often made by venture capitalists who finance developments so that they reach the market or by companies that buy developments to incorporate them into their production processes or as part of their products. In some specific cases, the commercialization of technologies is done through the sale of finished products to end-users.

In this sense, there are different reasons why someone might be willing to pay for a new technology in the market. One reason is to get ahead of the competition, that is, to generate competitive advantages through new technologies. Another is to access cutting-edge technology and knowledge without necessarily carrying out research activities, which are costly, long-term, and very risky. Companies are also looking to launch new products or acquire complementary industrial property, added to their own, as they can generate new products or services and market opportunities (INAPI, 2016).

There are two strategies for reaching a licensee. The first, called the "shotgun approach," and alludes to the image of launching many small projectiles, hoping that one or more shots hit the target. The most recommended mechanism for this a mass email blast, or some other mechanism of mass information distribution, which may seem effective since it has a wide reach. However, it is possible that considerable resources are required to implement this activity, or that people simply delete these emails without reading them.

The other strategy is called the "rifle approach." It means targeting a clearly defined audience effectively and efficiently through clearly defined promotional strategies. In this strategy, it is necessary to target specific industrial sectors or the right person within the company, such as the manager or vice president of business development or research and development. This strategy requires time and effort to identify the right people, but it has a higher probability of success. If no response has been received to the emails sent, it is advisable to send a reminder. If there is still no response, it is advisable to call to arrange a meeting to provide more information.

In either of the two strategies, it is important to provide non-confidential information, including a description of the features and benefits. It should not include the technical details of the technology, the patents filed, or the published patent applications, if any. In addition, it is recommended that relevant published scientific articles be included.

If the conversations progress and it is necessary to provide more information, it is recommended that confidentiality agreements be signed. It is never advisable to request the signing of a confidentiality agreement in the first or second conversation. Only do so when interest has been raised from the other party, and strictly confidential information must be provided, such as unpublished patent applications, unpublished articles, or a detailed scientific presentation.

To reach the target audience, it is necessary to address the appropriate audience and deliver the specific message of the correct industry. It is important, therefore, to carry out adequate market research, which is currently facilitated by multiple accessible and often free resources.

Market research can be contracted to specialized companies that perform this function. However, it can also access free information that is, information available on the web, such as scientific publications, patents and company websites. It can also access neutral third-party information, that is, without bias or particular interests, such as government databases or associations, universities and public laboratories, among others.

Regarding market research that can be carried out through intellectual property information, it is recommended to analyze the citations to verify if there are third-party patents where the technological development to be transferred has been cited. Without no registered patent, it can also check the closest matches to the developed technology.

Therefore, to commercialize a research result, it is very important to define the value proposition of the technology, identifying features and benefits; communicate the value proposition to potential customers in a simple, precise, and concise way; design a focused marketing campaign, identifying the correct target markets, key players, and contacts within the company, and, when appropriate, use market studies to analyze market size, detect opportunities, identify competitors and key players, and get to know its potential licensee.

References

Bauer, S., & Flagg, J. (2010). Technology transfer and technology transfer intermediaries. *Assistive Technology Outcomes and Benefits, 6*(1), 129–150.

Canetti, E. (2012). *Marketing of technologies* (p. 4).

Ibáñez, J. M. (2014). *Technology readiness levels: TRLS: An introduction* (pp. 165–171).

INAPI. (2016). *Technology transfer guidance guide* (p. 11). https://www.inapi.cl/docs/default-source/default-document-library/gu--a-gott.pdf?sfvrsn=48d557c2_0

MásQueNegocio. (2018). *Do you know the difference between marketing and advertising?* https://www.masquenegocio.com/2018/01/10/diferencia-marketing-publicidad/

NASA. (n.d.). Retrieved January 8, 2021, from https://www.nasa.gov/

Process of Technology Commercialization. (2016). Office of Technology Commercialization, The University of Texas at Austin. https://research.utexas.edu/otc/process-of-technology-commercialization/

Sullivan, N. F. (1995). *Technology transfer (Making the most of your intellectual property).*

Technology Transfer Offices

8

Abstract

This chapter addresses the crucial role of universities in technology transfer and explores the concept and importance of a Technology Transfer Office (TTO). The functions performed by a TTO are highlighted, and various support programs aimed at facilitating the transfer process, such as patent contests, early-stage financing, entrepreneurship programs, and international technology collaborations, are presented. Additionally, the chapter provides six fundamental principles for the success of a TTO. These include the commitment of authorities, the establishment of clear rules and regulations governing intellectual property and the creation of spin-offs, the provision of incentives, the development of effective strategies and structures, fostering a culture of technology transfer, and showcasing success stories. The chapter also discusses strategies for strengthening a TTO and emphasizes the importance of people as a crucial element in achieving successful technology transfer. Furthermore, performance metrics are explored to evaluate and measure the effectiveness of a TTO's activities.

Keywords

Technology transfer • Technology transfer office (TTO) • Early-stage financing • Entrepreneurship • Internationalization of technologies • Performance metrics

Role of Universities in Technology Transfer

Modern universities emerged in Europe during the Middle Ages. The University of Bologna, in Italy, is said to be the oldest in the world. However, although it was founded in 1088, it only received the title of university in 1317. The next institution

A. Ossa, *From Research to Market*, https://doi.org/10.1007/978-3-031-71340-8_8

to be created was the University of Oxford in 1096, which is recognized as the first English-speaking university. Universities are the oldest institutions and are among the few that have endured for centuries.

Today, universities have three missions. The first is teaching or the academic mission, the second is the mission to create new knowledge—also called the research mission—and the third is the mission to connect with society or the environment in which they are embedded. There is no consensus on the need for these three missions of universities, although this was not the case from their origins, at least not with this structure.

All were born with the teaching mission. In the seventeenth century, they began to publish in scientific journals, and from the last decade of the twentieth century, universities worldwide began to review their role within the knowledge society, formally incorporating the third task of relating to their environment, in addition to their traditional higher education and research functions.

This third mission also evolved. Currently, one of the means to fulfill this goal is to apply the results of their research work, that is, transferring knowledge to society, thus generating a new way of carrying out research, development, and innovation. Specifically, technology transfer allows the transmission of research results from those who create or own it to a third party in society or the market to solve a real problem in the environment.

There are some voices that argue that knowledge transfer or technology transfer is part of the second mission of universities since research does not only involve creating knowledge but also making it available. It is for this reason that technology transfer offices are, in most cases, organizationally structured under the umbrella of vice-rectorates for research or their equivalents and not from the vice-rectorates for community engagement. Be that as it may, the important thing is that technology transfer is a fundamental part of the work of universities, research centers, and, in general, those who create new knowledge.

From the perspective of the university world, it is also important to consider that technology transfer is relevant since it contributes to fulfilling the public role that these institutions have. In addition, it allows valuing research, that is, adding additional value to what they already provide through the training of people and the dissemination of knowledge.

On the other hand, technology transfer enables collaboration between universities and the public and private sectors, although one of its main focuses is to help solve multiple problems, contributing to the well-being of society and promoting the competitiveness of countries through the development of better goods and services. To achieve this purpose, universities have had to adapt and develop innovation strategies and more interrelated and flexible processes, as well as create support structures adaptable to R&D+i processes.

Finally, the evidence shows that to achieve these purposes, it is necessary to have a critical mass of high-quality scientific research (Wright et al., 2008). While the main research universities possess it, mid-tier universities still need to develop a broad and strong link with industry to drive innovation and technology transfer.

What and Why Is a Technology Transfer Office?

The history of the creation and strengthening of technology transfer offices dates back to the 1980s, as a result of a change in United States legislation. In December 1980, the Bayh-Dole Act was approved, named after the two congressmen who promoted it, with the aim of regulating inventions derived from federally funded research.

One of the fundamental changes made by this law was that institutions that had carried out the research could obtain the corresponding intellectual property when they had done so with state funding. Before this law, the commercial management of these results had to be managed by the government, and if institutions wanted to do so, they had to request express authorization. What happened before the enactment of the law was that the United States government had accumulated more than 28,000 patents, but less than 5% of these had commercial licenses; the law sought to reverse this (U.S. Government Accountability Office, 1995).

To achieve this, and in exchange for these greater freedoms, institutions were obliged to "grant the government a non-exclusive, non-transferable, irrevocable, and paid license to practice or have practiced for or on behalf of the United States the invention in question throughout the world" (*Public Law 96-571 96th Congress an Act United States of America in Congress Assembled, Civilian Energy Efficiency Swap Act of 1980*, 1980). In addition to submitting the patent application within one year after deciding to initiate its management, they should also report whether or not they would continue with it once the patent was requested, report on the use of inventions, and, to those who were granted exclusive licenses to use or sell in the United States, require that the products be substantially manufactured in the United States, among several other measures.

Therefore, the government decentralized control of inventions funded with federal funds, granting responsibility and authorization to commercialize them to institutions receiving state grants, with certain responsibilities to the government, the inventor, and society in general.

As a result, the creation of elements within institutions was required to fulfill all these requirements, called technology transfer offices or transfer and licensing offices (TTOs or TLOs).

The mission of a technology transfer office is to transfer research results to the market. They respond to the need to manage the transfer process formally, with clear incentives and policies, and generate long-term links with industry and the public sector. To accomplish this, it supports the research team in developing a strategy aimed at turning a research result with potential into a good or service that is available in the market and solves a problem in society.

To fulfill its mission, a technology transfer office must have a team of people—ideally, multidisciplinary—with complementary competencies to those of the research team; for example, in business, protection of research results, regulatory aspects, formation of technology-based ventures, relationship with industry, among others. Above all, technology transfer offices must work synergistically with their respective research teams.

According to the management and orientation of the office, there are two models. There is the internal level action (internalized offices), which works supporting the transfer and protection of technologies, and is part of the same institution where the research is carried out, in administrative and legal terms. Examples are the Massachusetts Institute of Technology, Stanford University, Cornell University, Columbia University, Austin Texas, Harvard University, most of the transfer offices in Latin America, and some in Europe and Asia. On the other hand, there is external level action (externalized offices) which operates for the institution to these offices they are linked to, but which are legally independent entities. These offices provide services to the institutions to which they belong and, in some cases, to other institutions and companies that require hiring their services in a particular way. Examples of this model are the universities of Oxford with Oxentia and the University of Cambridge with Cambridge Enterprise in the United Kingdom, or Queensland University with UniQuest in Australia.

Because embarking on the process of creating a technology transfer office is not easy, there are recommendations when undertaking this challenge.

The "Practical Guide for the Creation and Management of Technology Transfer Offices in Universities and Research Centers in Latin America" by WIPO states that "it is essential that the university verifies, beforehand and in a clear and precise manner if the indispensable objective conditions for the creation of a technology transfer office are met." In this sense, one of the necessary factors is to establish the existence or absence of research activities (Manderieux, 2011).

The guide states that, as a first step, the mission of the technology transfer office must be clearly defined and understood by all actors for it to be successfully developed. Then, the functions must be indicated according to the needs of its users. Finally, it is necessary to establish the tasks to be carried out and set the necessary institutional framework for the technology transfer office to adequately fulfill them.

Generally, the main users of a technology transfer office are the researchers at the university or research center. This is mainly because "researchers are the primary source of innovation that takes place within the university" (Manderieux, 2011).

However, the guide continues, there are other groups of users who could access the services offered by a technology transfer office, such as students who are developing their theses or research projects, private companies that require research and development services or support to identify and protect possible intellectual property assets, or public and private institutions, such as other universities, research institutes, independent inventors, among others (Manderieux, 2011).

Functions of a Technology Transfer Office

Technology transfer offices have the mission of transferring research results to the market. This mission includes making the best possible agreement between the university or research center and the company, where all parties benefit. To accomplish this, it is very important that transfer entities articulate conversations between actors who use different languages and have different incentives. In general, universities and scientific centers have a much more academic language, and their interests aim to impact society with the science they develop, without knowing much about how and with what implications that impact occurs beyond scientific publication. On the other hand, there is the business sector, which focuses on relevant areas for them, including legal or financial aspects. A very valuable dynamic exists when dealing with complementary views that allow achieving a joint objective, even when opposing views are often presented. The technology transfer offices must assume the fundamental role of uniting positions.

In addition to the above, it is necessary to generate networks (such as mentor networks) to access financial capital and various resources options that facilitate the marketing of goods and services that are developed within institutions.

Another fundamental role of every technology transfer office, and especially internalized offices, is to safeguard the interests of the institution or company they represent, for example, by ensuring compliance with applicable laws, current contracts, and internal regulations and policies of the institution or company. The process of converting an idea or research results into a product involves much effort and time and even requires skills that, in general, are not found in the research team, making it necessary to complement capabilities. In this sense, technology transfer offices should be a complement to research teams, especially in disciplinary capacities that they lack, such as the development of commercial strategies, support in legal aspects, or support in strategic intellectual property issues, among others.

However, it is important to consider that some ideas manage to reach the market and others do not, regardless of their academic potential. This is when technology transfer offices become relevant. In some cases, the offices accompany the process from the beginning of the research, and in others, from the time the research results have already been achieved, regardless of the stage of development they are in. After that, it is very important to validate whether or not there is a real need in the market or if the technology proposed is effective in solving the problem. It is also important to generate a preliminary transfer strategy, disseminate and protect research results, analyze market potential and competition, analyze the best business model, and validate research results. Also important is supporting the search for financing that is bringing the development closer to the market.

A cross-cutting element to the process, and part of the functions of every technology transfer office, is to generate a culture of innovation and entrepreneurship. This is done through various actions, such as training workshops and dissemination seminars, or by developing meeting spaces to include researchers, investors, entrepreneurs, and companies. In this sense, technology transfer offices become articulators with the different actors and organizations of the ecosystem since they have or should have contacts, as well as resources to facilitate interactions. All the above is extremely relevant because the lack of social capital is one of the most significant weaknesses when transferring technologies (Kantis & Angelelli, 2020).

The technology transfer office performs various functions, including managing public resources, creating science-based entrepreneurship, and fostering a culture supportive of innovation and entrepreneurship. The type and degree of support can vary, depending on the institution, from complete management of the previous processes to simple accompaniment and support for the research teams so that they themselves carry out these tasks.

Support Programs

One of the key roles of technology transfer offices is the coordination of ecosystem actors with activities that can contribute to the development of technologies. However, and depending on the reality of each country, some offices have their own programs that support scientists in these tasks.

- Patent contests

It is quite common for offices to have programs or activities that encourage patent applications (such as patent contests). It is also common to offer patent financing with the goal of supporting academics in the protection of intellectual property rights of inventions, to encourage the transfer to the productive sector and benefit society with these advances.

These contests offer to provide funds for or to carry out a state-of-the-art search of the proposed inventions, provide advice, and offer the support of a specialized team in the protection strategy and in obtaining funds to continue the patent and technology transfer process. Additionally, assistance is provided to draft and submit the patent, that is, financing for the drafting of those inventions whose prior art searches indicate that they are patentable.

It is important to indicate in these contests the obligations that academics must assume. For example, the winners much actively participate in the protection process, respond to inquiries made by office members for the development of the prior art study, attend coordination meetings with experts, provide information for the drafting, and review the drafts of the patent, among other obligations.

It is also relevant to explicitly state in this type of contest the selection criteria for the winners, such as the relevance of the problem solved by the invention, the expected results after sustaining the patent, and the level of development of the invention's research, taking into account elements such as prototype, laboratory tests, in vitro tests, in vivo tests, final product, potential impact generated, and characteristics of the research team. In some cases, these contests are focused on a specific protection mechanism (e.g., copyright or industrial design), but sometimes, intellectual protection contests are held so that research results are submitted regardless of the mechanism under which it is protected.

As part of the selection process, it is highly recommended that the presented technologies be validated with industry members, thus allowing for adjustments to the projects and eventual patent applications in advance. This helps to avoid investing time and resources in technologies that do not meet market needs.

This type of program contributes to one of the main functions of technology transfer offices, which is the search for research results or "scouting" of technologies.

Channeling this type of funding through contests is recommended because it makes the process more competitive, allows evaluation by third parties, and, above all, is relevant because researchers are used to participating in peer evaluations.

- Early funding

Other interesting programs to develop are early funding, or "Gap programs." Their objective is to cover the gap that exists between laboratory development and the market, so they seek to accelerate the transfer and commercialization of new technologies and inventions developed by researchers from various disciplines of the university or research center.

These programs are aimed at researchers and academics who have a project with some progress in the results of applied research, such as laboratory-scale proof of concept or small-scale prototype carried out in the laboratory.

This type of program aims to allocate resources to finance prototype testing activities in a real environment to reach a functioning pre-commercial product or, ideally, a first version ready for marketing. In general, amounts between $20,000 and $100,000 are financed per project, and between three and eight projects are supported per contest depending on budget availability and the R&D capacity of the institution.

Projects suitable for funding through this type of contest have generally already progressed on the path to obtaining more traditional research funding, but have not yet reached commercial viability. An example would be a case where viability has been demonstrated by an academic prototype but funds are required to develop a commercial-grade prototype.

This type of program has been shown to have a positive impact on technology commercialization and support for the creation of new science-based ventures.

- Entrepreneurship program

It is very common to find entrepreneurship programs within higher education institutions. However, science-based ventures have particularities that make the programs and support different.

The objective of this type of program should be to add economic and social value to the country and the world, promoting science and technology-based projects towards stages closer to the market. It generally supports the prototyping, business development, and packaging phase, pushing the evolution of a basic prototype to an advanced one to streamline the venture's progress. This requires that the projects have laboratory results, proof of concept, or prototypes that demonstrate their technical feasibility.

In this type of contest, teams are generally asked to present a scientific-technological basis for the project, information on the problem they solve, and a description of the proposed solution. In addition, it is important to know the level of development, that is, if the results are in the state of experimental tests, proof of concept, or prototypes, clearly detailing the market and commercialization opportunity. One of the crucial factors for obtaining funding is the capacity of the team to carry it out, so it is necessary to provide a detailed description of its members. Since the creation of science-based ventures requires financial resources, some leading international institutions have advanced in establishing their own pre-seed funds or in partnership with other universities (Kantis & Angelelli, 2020). While financing alone does not work, it is necessary to generate a permanent support capacity through entrepreneurship programs or academic entrepreneurship programs, which help to select those projects with the greatest potential and interest, contribute to working hand-in-hand with the development of the value proposition and business model, and favor obtaining financing for their takeoff. At this point, it is very important to understand that a venture can be strongly supported, but it only moves forward if the researcher in charge is convinced and truly determined to lead this challenge.

- Internationalization of technologies

Another relevant program is the internationalization of technologies; this makes a lot of sense, especially in those countries with limited or non-existent markets. Its objective is for researchers to introduce their applied research results into global markets. This type of program aims to generate entrepreneurship and commercialization capabilities in research results. In addition, it seeks to link the developers of these types of projects with brokers of technology, companies, or other international research centers to bring these technologies to global markets.

One of the fundamental objectives of these programs is to generate entrepreneurship and commercialization capabilities in researchers. It is therefore suggested that the applicant teams be composed of two people: the principal investigator of the team and someone interested in the commercial aspects related to the massification of the technology.

Six Golden Rules for the Success of a Technology Transfer Office

The creation of a technology transfer office is not an easy task. The following are the critical factors that give a higher probability of success in this path, a proposal that we collect under the name "Six golden rules for the success of a technology transfer office" [1].

1. The commitment of the authorities
2. Clear regulations
3. Incentives
4. Strategy and structure
5. Culture
6. Success stories.

The Commitment of the Authorities

One of the most critical factors for the success of any technology transfer office is the commitment of the institution's authorities, guarantors of the innovation process. It is important that the institution's mission recognizes the fundamental role of technology transfer, which must be transmitted from the highest authorities to each academic unit to each employee, student, academic, and member of the institution's community.

Involving private companies and public institutions in the technology transfer process while presenting the institution as a strategic partner with benefits for their own development and that of the country is another task in which the authorities have a relevant participation.

The authorities play a key role in highlighting the importance of supporting the work of researchers in transferring their projects, considering their allies in the process, as well as their academic units. Because of this, they work closely with the researchers to define strategic improvements in innovation, transfer, and protection in order to increase the impact of research.

This commitment of the authorities allows the transversal articulation of the entire process. It impacts academically on undergraduate and graduate students, and on the work of teachers and researchers, promoting interdisciplinarity, entrepreneurship, and innovation. Additionally, it facilitates the internationalization of the university both with the research projects of students and academics presented, and in collaboration, with foreign universities.

For the authorities, technology transfer activities contribute to improving the institution's reputation, and allow citizens and the government to see that the organization is dealing with real problems in society, and actively participating in economic, social, and cultural development.

Clear Rules

Another key element for the success of a technology transfer office is to have clear rules regarding the technology commercialization process, both for researchers and for the industry that wants to connect with the university or research center. To that end, rules are established that regulate, indicate restrictions, and provide incentives to facilitate the process.

These rules must address aspects such as the compatibility of the researcher's academic and entrepreneurial tasks, as well as participation in intellectual property and the share package of a potential new company.

These regulations may vary depending on the reality of each country, but they usually have common aspects around the world.

Intellectual Property Regulation

In every organization that commercializes technologies, regulations that control intellectual property must exist. These aim to regulate the rights and obligations regarding works and inventions developed by those participating in research activities.

The regulations must include the duties of both the institution and the researchers who participate in the process. Additionally they must regulate aspects such as the ownership of research results, the duty of communication of developments by researchers, the significant use of means or resources provided by the institution, financial compensation for the parties, and the obligation of confidentiality.

When the creation has been the product of the work of a person specifically hired by the institution, research results belong to the institution. They also belong to the institution in these cases: when the research has been carried out by a person external to the institution where there is a prior agreement in which the assignment of intellectual property rights is contemplated; when it has been carried out in whole or in part with significant use of means provided by the institution; and, depending on the laws of each country, when it involves computer programs developed or produced in the performance of work functions.

In the case of intellectual creations such as literary, artistic, and scientific works, regardless of their form of expression, although ownership may belong to the institution, it must commit to safeguarding the rights acquired by the authors of works, obliging itself to recognize the moral right included in copyright, which by law always belongs to the author or creator of a work. In addition, the institution, assumes as an obligation the express recognition of the author's name in any type of internal or external communication to it. First of all, it is necessary to verify what national legislation establishes in this regard.

Where research results are protected through some mechanism of industrial property law, these belong to the institution when the inventor is a person hired in a dependent or independent relationship by it, or when the invention has been developed by a person not hired by the institution but who has made significant use of means provided by it (e.g., students).

Regarding the duty to communicate, the regulations establish that those who carry out any creation, discovery, or invention with the potential to be patentable or intellectually protectable must communicate it as soon as possible to the corresponding body within the institution, in writing, with the utmost discretion, and prior to any publication or dissemination of the work.

One of the key aspects within an intellectual property regulation is the definition of "significant use of means." It is to be considered that there is significant use of means provided by the institution when resources such as specialized equipment and materials for research purposes, the name, emblem, brand, or representative images of the institution, and the services of the institution's officials are used.

The use of office materials and equipment, including software for general use, photocopiers, personal computers, documentation, and material contained in the library or other resources that are frequently available outside the institution, is not considered significant use of means provided by the institution.

When there is controversy over whether or not significant resources were used to obtain a research result, it is the unit in charge of technology transfer or the highest authority in the area to which the research team belongs, who determine it. If there is no agreement, commissions that can analyze and resolve the conflict should be formed.

Commercial compensation is another fundamental aspect of these regulations. Agreements change depending on the country or even depending on the institution within a country. In some cases, it is established that 15% of the economic benefits that the institution perceives as a product of the commercialization of intellectual property rights go to the technology transfer office that takes charge of the processing and management of intellectual property and commercialization. Of the remainder, one-third is distributed to the research team, one-third to the faculty or department from which the research team comes, and one-third to the institution. In some cases, larger amounts are established for the research team as a way to encourage researchers to innovate. When transferring technology in a university setting, it is essential that the university reserves the right to use the transferred results for educational, teaching, and research purposes.

Confidentiality obligations are also very common in these regulations. They establish that research teams have the obligation to keep confidential any information that could affect obtaining any intellectual property rights, and that the teams must sign the respective confidentiality agreements.

Finally, these regulations also determine how and who resolve potential conflicts that may arise. It is often determined that the first instance is handled by some competent authority, and any instances thereafter is handled by a specialized internal committee, formed for such purposes.

Spin-Off Creation Regulation

Another very relevant regulation is the one for the creation of science-based ventures (as detailed in Chap. 5), also called "spin-off regulations." This type of regulation establishes the duties and rights of both the institution and the researchers themselves and enables researchers to participate in the transfer of research results through the creation of spin-offs.

This regulation clearly establishes the approval process conditions for creating a spin-off within an institution, as well as certain incentives for research teams to create them. It also regulates the relationship that must exist between the institution and the new venture, as well as the potential shareholding participation of the institution in the spin-off. It is also advisable to regulate the possible participation of officials and other people hired by the institution in the activities of the spin-off, at least in its genesis and in relevant steps. If not regulated, they can suggest future conflicts—a factor that can inhibit the interest of an investor.

Regulations must be clear as to whether or not members of the research team can be shareholders of the spin-off. Additionally, the potential conflict of interest that is generated by being both a member of the licensing institution and the licensee company must be regulated. This is made possible by establishing, for example, whether they are able to participate in and know about the negotiations of the license with the institution. However, if this is allowed, direct negotiations regarding their terms must be prevented. Regulations also determine whether or not people under contract with the institution may hold the position of manager or president of the new company or if they can hold only the position of scientific director or consultant. Potential conflicts of interest must always be regulated.

Conflicts of interest that may arise as a result of the transfer of research results and these must also be regulated. It is very important to note that the conflict of interest per se is not the problem; the problem is when it is not declared and not properly resolved.

It is understood that a conflict of interest occurs when a member of the institution has (or may have) the opportunity to exert influence on the decisions that the institution adopts within its scope and that these may imply or allow personal gain or advantage, of a relative or related persons, and when a member of the institution has directly or indirectly interests that affect (or may affect) their ability to make independent decisions in the fulfillment of their functions with the institution.

Conflict of Interest Regulation

The regulation governing conflicts of interest is crucial in the technology transfer process. Essentially, this regulation establishes that members of the institution are obliged to always act with integrity in their functions and avoid obtaining inappropriate personal benefits from them.

Although it is true that conflict of interest in and of itself is not bad, the aim of these regulations that they are quickly identified, declared, and resolved in the best possible way.

As we have pointed out, there is a conflict of interest when someone has the opportunity to exert influence on decisions made by the institution, which may generate personal advantages, or when they have direct or indirect interests that affect their independence in making decisions. This can often happen when a spin off is created, whether or not the institution participates, and certain agreements are generated between both instances. Examples include, but are not limited to: influencing the purchase of goods that benefit the academic or the spin off; receiving gifts or special favors with the aim of influencing the institution's decision-making using the institution's resources for their own benefit or that of their company; using confidential information for personal gain; or using infrastructure or equipment belonging to the institution without authorization.

The regulation is the procedure for resolving the potential conflict of interest, determining who should be informed, when, how much time there is to analyze it, and the type of measures to be taken.

Among the measures adopted are the prohibition or authorization to carry out the activity that caused the conflict or the authorization of said activity subject to compliance with certain conditions that seek to eliminate the negative effects of the conflict.

In the procedure, establishing an appeal process, as well as sanctions for non-compliance with the resolution, are suggested.

While it is true that the conflict of interest regulation is very useful for the field of technology transfer, it also covers many aspects of the relationship between the researcher and the institution.

Incentives

It is very important to understand that researchers are the protagonists of the technology transfer process, so it is key that institutions focus much of their efforts on creating research and development capacities, both material and human, as well as generating the necessary incentives for researchers to become involved in technology transfer aspects.

Funding for research is an important incentive. In most countries around the world, research funding comes mainly from government agencies through competitive funds or so-called baseline funds, which in most cases are also competitive. Working together with the private sector is also key when it comes to financing science, as is funding from international sources. Therefore, it is important that institutions can articulate and facilitate the process of raising both public and private resources and that, as far as possible, the institution makes its own funds available for research projects through internal competitions.

It is also appropriate to encourage and generate interdisciplinary research that allows an approach to problems from different perspectives. This facilitates finding innovative solutions and can be done by creating competitions that have a higher probability of being awarded if the team is composed of researchers from different disciplines.

Another way to create incentives for research production and technology transfer is the generation of internal regulations (such as those regarding intellectual property and the creation of spin offs) that clearly define the processes and benefits for researchers involved in technology transfer activities. Within the aspects governed by these regulations are financial incentives, where the way in which potential income received by the institution for these concepts is distributed among the researcher(s) is established. In this sense, the most fertile universities in terms of technology transfer and spin-offs are those that grant inventors an adequate share in the commercialization results (Bramwell & Wolfe, 2008).

Incentives can also be provided through the creation of "protected time" for research and technology transfer, such as a sabbatical period, or a reduction in other types of duties, such as academic ones. Since it is necessary to allocate time to validate a good or service at a technical and commercial level, develop the team and contacts, and access and organize resources that allow for the realization of innovation and potential entrepreneurship, it is important to have rules and regulations that allow the researcher to reconcile these activities with their academic career (Kantis & Angelelli, 2020).

In the case of creating spin-offs, some institutions have established an incentive whereby the institution may license the research results under more favorable conditions to the spin-off than to third parties, such as granting a first option for licensing, more beneficial royalty payment conditions, longer license duration, etc. Some institutions have also allowed the researcher participating in a spin-off to request a leave of absence or modification of working hours, with the right to reserve their job position, either without salary or with the respective salary adjustment, and without loss of seniority. The use of physical space has also been considered. In the case of potential income, the spin-off shareholders may withdraw profits when appropriate and, at the same time, receive royalties for previously developed research results.

Another key aspect is to establish rules that take into account the innovation performance of researchers, recognizing and incentivizing their academic careers. That is, universities should establish clear promotion incentives in the academic career for those researchers who are involved in technology transfer and innovation activities.

Prestige and recognition are other incentives for researchers involved in technology transfer matters. For example, an annual technology transfer ceremony can be organized, inviting different actors from the academic, business, and government worlds related to these matters, and awards given to academics who have obtained patents, to those whose results have been licensed, and to faculties that historically have the highest number of patent applications, among other categories.

Incentives are very relevant, as university academics are not required or mandated to follow this path; in fact, international figures indicate that around 20% of university researchers participate in technology transfer activities.

Strategy and Structure

The basic concept of strategic planning states that a strategy cannot be carried out without organizational structure and that, in turn, an organizational structure cannot be conceived without a strategy. For larger organizations, the relationship might be more complex than for a smaller organization, although it is very important that the organizational structure is developed and evaluated as the strategy evolves, so as not to succumb the temptation of restructuring according to the situation, but rather according to what is pursued.

In the case of technology transfer offices, it is very important that their strategy is aligned with the institution's strategy. In this sense, it is advisable to carry out a thorough analysis of the strategic plan, then verify how the office's actions contribute and report to complete said plan. These tasks are not considered as a parallel and additional activity, but as an activity that is completely aligned with the institution's pursuits.

It is recommended that the strategic plans of the offices be medium- and long-term (five or more years) and that they have operational plans for one or two years at most. It is also important that this plan includes interaction with the different key actors of the institution's entrepreneurship and innovation ecosystems, such as incubators, accelerators, liaison offices, or entrepreneurship centers since they all have innovation, product development, and their bridge function between academia and society in common. However, it is essential that each of the organizations is clear about where their role begins and ends and that the human resources profiles they have are appropriate.

It is essential that the structure adapts to the culture of the organization and that the profiles are suited to the needs of positioning, intellectual property management, and technology commercialization, at a minimum. In some cases, the structures respond to a compartmentalized organization according to functions; that is, on the one hand, intellectual property is managed, and, on the other, technology commercialization. This compartmentalization is advantageous when specializing functions, although it poses the challenge of having relevant articulation that allows some actions to move from one structure to another or that ultimately no area takes charge of actions that it believes do not correspond to it. Another mechanism is to have a matrix structure where the areas are responsible for the execution of activities in their entirety, although they use capacities from other functional areas of the office. In this type of structure, work is much more integrated, but strong coordination can be a difficult challenge, since at times, a person may feel they have two bosses.

Regardless of which structure is adapted, the important thing is that it is aligned with the strategy and that it is reported to the institution's development plan.

Technology Transfer Culture

The creation of a technology transfer culture within the institution is essential to strengthen the ecosystem and includes workshops, seminars, courses integrated into the curricula of undergraduate and graduate programs, innovation promotion activities, and contests.

The culture of innovation plays a central role in the evolutionary capacity and competitiveness of universities. Making the knowledge of this culture explicit allows institutions to connect each of the members of the institution as key actors and participants in its development. Strategically, institutions, must establish actions aimed at making innovation and technology transfer play a fundamental role within university activity.

Among the activities that can be carried out are a recognition day, or an innovation or technology transfer day—an opportunity in which the work of academics and students in this area is rewarded. Holding it once a year, through a formal and structured ceremony, and with the participation of the highest authorities is recommended.

Other activities are cycles of talks on specific subjects, such as intellectual property, technology transfer, or the generation of spin-offs, which may be aimed at researchers, doctoral students, and the university community in general. These talks are informative, short in duration, given either by experts from the transfer office or external to the institution, and carried out with ample time to answer questions. Depending on the culture of the organization, an alternative is to hold talks for specific audiences (i.e., not necessarily open to everyone), and also some aimed at the general public so that the different actors feel included.

It is also important to generate thematic dissemination activities, e.g., according to research areas (education, biomedicine, etc., or the one corresponding to the institution. Since each of them has particular specificities, it is necessary to disseminate methodologies and innovations in their respective fields and disciplines.

Collaborative work sessions also promote the transfer culture, with the aim of training professionals and technology managers from different areas of the institution on technology transfer, intellectual protection, and science-based entrepreneurship topics.

The implementation of elective courses on technology transfer and intellectual property, aimed at both undergraduate and graduate students, also promotes a technology transfer culture within a university. Teaching them about science-based entrepreneurship and provide practical experience is very relevant since they are some of the very people who one day develop successful activities. Ideally, these activities should be practical workshops rather than theoretical ones, so that they develop the necessary academic training with real projects.

A plan that includes activities (e.g., seminars, congresses, and fairs) that promote innovation can position the institution at the national and international levels of innovative, science-based production. Additionally, interactive web platforms that allow dissemination and that highlight the portfolio of projects to be transferred, innovation activities, applied research, or technology protection bring external attention to the institution while internally reinforcing and strengthening the institution's technology transfer culture.

Success Stories

According to published statistics (Stevens & Burley, 2003) referring to new business development processes, out of 3000 ideas, only one becomes successful in the market. Of that total, 300 were presented to someone and 125 of them materialized into a formulated project. Of the 125 projects, as can be inferred from the report, nine move on to an initial development phase, so four of them reach a level of relevant importance. Finally, only 1.7 becomes a good or a service that can be launched into the market for only one to be successful.

These discouraging data reveal that innovation presents high risks and has very low success rates. Therefore, it is very important to demonstrate that all the effort dedicated to technology transfer is relevant, by obtaining some success story in the short term.

This is not easy and can almost become a gamble as chance intervenes; the chances of winning or losing do not depend exclusively on the player's skill.

Therefore, in the initial stage of establishing a technology transfer office, it is very important, in addition to many other variables, to focus on obtaining some success in the short term. This can be achieved by selecting a group of promising projects and being flexible to achieve the transfer. It is possible that the first cases do not involve significant economic returns, and it is also possible that concessions and adaptations to circumstances is necessary to achieve it. The ultimate relevance is to bring a research result to the market to demonstrate that the goal can be achieved.

As Isaac Newton stated in his first law, "A body cannot change its initial state by itself, whether at rest or in uniform rectilinear motion unless a force or a series of forces with a non-zero resultant is applied" (*Newton: Leges Motus from the Principia*, n.d.). This inertia must be broken, and the necessary energy must be used to change the initial state of technology commercialization within an institution. When this occurs, the second, third, and nth transfers are much easier or much more beneficial.

Strategies for Strengthening a Technology Transfer Office

Based on the experience of the Pontifical Catholic University of Chile, and according to its Technology Transfer Manual, the following are different strategies that strengthen technology transfer offices within an institution.

First, promote specialization and autonomy in the organizational structure by reinforcing the capacities that aim to achieve relevant knowledge related to innovation development and transfer in team members. Team members must have competitive intelligence which, among other things, help them understand how to detect transfer opportunities based on research results, negotiate transfer contracts, and protect strategies for research results.

Second, streamline the management of support for technology transfer based on a review of the work model and transfer methodology carried out by the transfer office for the management of project portfolios, proposing permanent improvements in processes that allow a more expeditious flow for the transfer.

Third, there should be stable resources for the operation of the office, mainly for the team of people. The institution must focus on attracting base funding, complementary to institutional funding, that allows it to consolidate, improving the quantity and quality of its activities. This means generating a financing model that provides a diversified resource matrix for the office's long-term growth, which includes internal funds from the institution, funds from public competitions, both national and international, and private financing.

Fourth, private financing for technological developments supported by the office should be strengthened. Obtaining financing from private sources to implement research and development projects is essential for the success and sustainability of the technology transfer office and the projects it supports. These projects contribute to the development of technologies that are applied and transferred to the market and society.

Fifth, create links with venture capital networks. Venture capitalists seek to invest in technologies, betting on technological solutions to global problems, focusing on large industries, and minimizing risks in a diversified investment portfolio. They invest in companies with developed innovation, disruptive potential, and high competitive advantages, aiming to generate high returns. This connection helps to bridge the gap in the financing chain for science and technology-based ventures that are not yet covered, and it can be an attractive investment opportunity and access to innovative projects, supporting the development of ventures on a global scale.

Sixth, periodically review the regulation of processes supporting technology transfer. It is necessary to revisit intellectual property regulations, the creation of spin-offs and others, at least once every ten years. Because the ecosystem is changing, it is necessary for the rules to align with incentives and any changes to national laws.

Seventh, strengthening the existing infrastructure that contributes to the development of research and development projects is another relevant strategy for strengthening technology transfer activity within an institution.

Eighth, strengthen the generation of contact networks and participate in associations or groups specialized in intellectual property and technology transfer issues, as this allows the sharing of best practices and learning about experiences and problems that others face, which are usually very similar between one organization and another.

People Are an Essential Element for Achieving Technology Transfer

The main asset of a technology transfer office is its people. In this sense, they are key to achieving successful technology commercialization. The document "Maximizing the impact of your TTO" (Bos, 2017) recommends that to ensure a competent team of people in a technology transfer office, they have the "ability to relate and understand academics and their academic work," that is, they possess scientific knowledge to understand developments in general terms and to facilitate dialogue with researchers.

The document states that it is essential to have excellent communication, relationship, and teamwork skills, which are key when managing technology transfer. It is also important to have to have business experience to understand the business world and its needs. Technical knowledge is also highly desired, as it allows dialogue with researchers, as well as a thorough understanding of what is to be commercialized.

According to the author, it is necessary to have the ability to manage many projects simultaneously and to prioritize work. Prioritization is essential for managing a technology portfolio.

The work team's social capital is highly desired (Bos, 2017). It allows the construction of contact networks—a key element when linking the university and companies –and is essential for commercializing technologies.

Clearly, having people with all these characteristics and skills is not easy, so when starting recruitment, it is recommended to check the office's needs and what knowledge candidates would need to learn or acquire. This allows the development of training and mentoring programs for new skills and professional development to have the necessary capabilities to carry out the work (Bos, 2017).

The work team's experience is essential—not only the experience they bring when forming the teams but also the experience they acquire working with the team over time. For this reason, it is crucial to avoid high turnover in technology transfer teams and it is important to generate incentives to retain them. Those offices that achieve low turnover in key positions better capitalize on the intellectual capital they have developed and, with that, increase the success rates of technology transfer office management.

Performance Metrics

To evaluate the performance of technology transfer in an institution and, therefore, the results of a technology transfer office, it is essential to have performance indicators. The University of British Columbia, has established a set of metrics and divided it into four impact groups: academic impact, economic impact, financial impact, and social impact.

Regarding academic impact, four evaluation criteria are proposed: the number of students trained during the development of the technology; the number of publications based on the technology and impact assessment of journals; the level of funding for research and development of the technology; and the ability to continue supporting future developments alongside the academic progress of the researcher. Academic impact is the most common indicator, with the greatest potential, and is often measured or perceived at the time of technology transfer, demonstrating that technology commercialization is not at the expense of academic development.

Economic impact is one of the most obvious criteria of technology transfer management. Within it are the number of people hired in spin-off companies, the total sales of the spin-off companies, and the sales generated from technologies licensed to third parties. The University of British Columbia´s approach states that half of the licensed technologies reach their potential impact after approximately five years, that most licenses have a low economic impact, and a small number have a very significant impact.

Financial impact measures the level of annual royalties or other payments received by the institution according to the license agreement, as well as the value of shares in the spin-off companies. It is important to understand that most licensing revenue comes from a small number of licenses and that technology licensing is not primarily motivated by potential financial returns but by the impact in other areas.

In the social impact, the criteria are the relevance of the topic being addressed, for example, diseases, environmental issues, or minority group issues. Another criterion is the prevalence of the problem faced and the role of the technology in reducing the threat or prevalence of the problem. The university states that for licensed technologies whose greatest potential impact is social, half of these licenses take fifteen to twenty years to reach their potential, and medical therapy-related technologies often offer the greatest potential impact of their portfolio.

Another relevant aspect raised by the University of British Columbia is that the licensed technologies that form the basis of spin-off companies have higher actual and potential impacts overall compared to other licenses. For example, it is very likely that large financial returns for an institution come from equity participation in science-based ventures.

Other metrics can also be used to evaluate technology transfer activities in knowledge-creating institutions. These include networking and training, proposed by Praxis Unico, *Metrics for the Evaluation of Knowledge Transfer Activities at Universities* (Holi et al., 2008), measured by the number of people who met at events organized by the transfer office that led to innovation activities. To gauge the efficiency of these activities, the percentage of sustained events that led to transfer activities can be estimated. Another metric is the development of professional training activities, which can be measured by the number of courses developed, and people and companies attending. To measure the quality of these activities, there are indicators such as feedback and evaluation from attendees.

Regarding the consultancies that a technology transfer office could provide, one can measure the number and level of income from contracts, the percentage of income relative to other income, market share, the number of clients, and the length of the relationship with clients. The measure of their quality is based on the percentage of repurchase, evaluation, and feedback from clients and the quality and importance of the client.

Performance can also be evaluated based on collaborative research and contract research agreements. These can be measured by the number and value or income of contracts, the percentage of income relative to total research income, market share, or the length of the relationship with clients. To gauge the quality of the activity, the percentage of repurchase, evaluation and feedback from clients and the number of products successfully created from the research can be measured.

Another indicator is patents. These are measured according to the number of patent applications, such as priorities, international applications, etc. However, it is essential to understand that patents are a means and not an end in themselves.

Licensing is one of the key indicators of any technology transfer office. This is measured by the number of signed licenses, the revenue generated through licensing, or the number of products generated from licenses. To measure the quality of these licenses, there is customer evaluation, the quality of the licensee company, and the percentage of revenue generated through licensing versus other mechanisms for generating income from technology commercialization.

Finally, regarding the creation of science-based ventures or spin-offs, the number of companies created, the sales generated, the external investment raised, the market value of the companies, the survival rate, the quality of the investors, investor-customer satisfaction, and the growth rate of the ventures can all be used as performance indicators.

Therefore, contrary to what one might think, the financial aspect is only one element and not necessarily the most revealing in determining the impact of technology transfer. Elements such as improving the reputation of the university or scientific center where the development occurs and enhancing national and international credibility for the institutions where these developments occur are fundamental indicators.

Whatever metrics are used, the impact of technology transfer must be measured exhaustively, considering the multiple factors related to this activity.

Note

1. Own creation ↑.

References

Bos, A. (2017). *Maximising the impact of your TTO introducing the trainer*.

Bramwell, A., & Wolfe, D. A. (2008). Universities and regional economic development: The entrepreneurial University of Waterloo. *Research Policy, 37*(8), 1175–1187. https://doi.org/10.1016/j.respol.2008.04.016

Holi, M. T., Wickramasinghe, R., & Van Leeuwen, M. (2008). *Metrics for the evaluation of knowledge transfer activities at universities* (Vol. 5, pp. 1–33). Library House, Cambridge. https://ktdrr.org/ktstrategies/ktstrategies_search.cgi?location=sr&sel_1=109&pedisable=true

Kantis, H., & Angelelli, P. (2020). Science and technology-based ventures in Latin America: Importance, challenges, and recommendations for the future. In *Science and technology-based ventures in Latin America: Importance, challenges, and recommendations for the future*. Inter-American Development Bank. https://doi.org/10.18235/0002156

Manderieux, L. (2011). *Practical guide for the creation and management of technology transfer offices in universities and research centers in Latin America*.

Newton: Leges Motus from the Principia. (n.d.). Retrieved January 8, 2021, from http://www.thelatinlibrary.com/newton.leges.html

Public Law 96-571 96th Congress an Act United States of America in Congress Assembled, Civilian Energy Efficiency Swap Act of 1980. (1980). (p. 4).

Stevens, G. A., & Burley, J. (2003). Piloting the rocket of radical innovation. *Research-Technology Management, 46*(2), 16–25. https://doi.org/10.1080/08956308.2003.11671550

U.S. Government Accountability Office. (1995). *United States General Accounting Office GAO Report to Congressional Requesters* (Issue HEHS-95-61).

Wright, M., Clarysse, B., Lockett, A., & Knockaert, M. (2008). Mid-range universities' linkages with industry: Knowledge types and the role of intermediaries. *Research Policy, 37*(8), 1205–1223. https://doi.org/10.1016/j.respol.2008.04.021